PROJECT MANAGEMENT FOR THE 21st CENTURY

SECOND EDITION

Praise for the First Edition

"Focusing on proven methods to deal with the economic climate and technologies currently facing managers, this text discusses how to employ technology effectively, covers organizational and behavioral aspects of project management, provides advice for dealing with 100 common problems, and serves as a practical guide to setting up and managing projects."

—*PMI Information SourceGuide*

"The authors have tremendous insight into the dynamics of teams and projects; they are strong in practical detail, communication, and ways to solve sticky problems. In presenting material never before available to the industry, they have created a welcome addition to any library."

—*Sudir Jain, Integrated Management Systems, Inc. (IMSI),*
Ann Arbor, Michigan

About the Authors

Bennet Lientz has taught and consulted on project management for the past 28 years. Kathryn Rea has consulted on project management with over 50 major firms and governments in Europe, North America, Asia, and Latin America. Both have delivered seminars on project management to more than 6000 students on five continents.

PROJECT MANAGEMENT FOR THE 21st CENTURY

SECOND EDITION

Bennet P. Lientz
University of California, Los Angeles
Los Angeles, California

Kathryn P. Rea
The Consulting Edge, Inc.
Beverly Hills, California

Academic Press
San Diego London Boston New York Sydney Tokyo Toronto

Academic Press
a division of Harcourt Brace & Company
525 B Street, Suite 1900, San Diego, California 92101-4495, USA
http://www.apnet.com

Academic Press Limited
24-28 Oval Road, London NW1 7DX, UK
http://www.hbuk.co.uk/ap/

Library of Congress Card Catalog Number: 97-80797

International Standard Book Number: 0-12-449966-X

PRINTED IN THE UNITED STATES OF AMERICA
98 99 00 01 02 03 EB 9 8 7 6 5 4 3 2 1

Contents

6   Being a Winning Project Manager

7   The Project Team

8 Intranets, Internet, and the Web

III MANAGING THE PROJECT

9 Tracking and Monitoring the Project

10 Project Cost Analysis

15 Project Management Software

16 Project Administration

V EVOLUTION, REVOLUTION, AND TERMINATION

17 Project Change and Death

18   Lessons Learned from the Project

19   Where to Go from Here

Preface

Rightsizing, reengineering, global alliances and relationships, and technology advantage are trends in the last years of the 20th century and the first decade of the 21st century. What do these disparate topics have in common? To achieve success, they must be managed as projects. How does today's environment differ from the past? There are fewer constants, more challenges, reduced resources, and increased risks—factors that support project management. Project management has risen to become the major management technique for addressing most major work.

Traditional project management has been successful in achieving specific objectives. It has been expanded over the years to address organizational and managerial as well as technical situations. Properly applied without massive administrative overhead, project management supplies structure, focus, and control to drive a project team to success. What has been needed is the modernization of project management to reflect both the wide range of projects and the new technology available.

Project management itself has not been immune from technology. From single-user project management software we now have network-based project management, enabling the sharing and visibility of project information and supporting a team approach to project management. The range of tools useful to project teams has grown to include electronic mail, the Internet, videoconferencing through the Internet, electronic forms, intranets, database management systems, and groupware. Anyone attempting to compete and succeed without these modern tools will be handicapped and their risk of failure will increase. Why? Because these tools work among people. They help people work together to resolve issues, obtain status, and share a common vision of the future.

Project management has been transformed through technology in terms of the scope of projects. It was much more difficult in the past to manage cross-departmental, global, cross-industry, and large-scale integrated projects within firms. From having a single-person focus, project management has been extended to the project team and beyond. These trends are reflected in the second edition, which addresses collaborative project management, group resource deconflicting and allocation, and information sharing. These are also reasons for the growing spread of project management.

We are very grateful for the response and suggestions of readers of the first edition. This second edition expands the guidelines and the use of modern technology and spends more time on project analysis and costing and issue management. Specific topics that have been expanded include:

- Using the World Wide Web and Internet in project management and projects
- Groupware, database management, videoconferencing, and other network-based technology for information sharing, tracking, and issue management
- Project-oriented intranets within and across companies

Why did we write this book when there are many others on the market? All of the books provide definitions and basic concepts of project management. Few provide tangible guidelines on how to do project management, how to avoid failure in projects, and how to increase the odds of success. Most do not address new technology, but instead are time-locked in the 1970s or early 1980s. Many are very dry and academic. Our approach is different. We provide:

- Specific and detailed guidelines for help with projects
- A down-to-earth writing style
- Real-world examples of success and failure

- A wider scope covering all project activities—from concept to completion or death
- Directions on how to get the most out of modern technology

The organization of the book reflects our approach:

- Understanding the changes and trends in projects (Part I)
- Defining and setting up a successful project plan (Part II)
- Managing the project (Part III)
- Selecting and using modern project management methods and tools (Part IV)
- Dealing with project change, ending a project, and what to do next (Part V)

Each chapter contains guidelines and steps to take next. We also provide an integrating modern project example that spans the chapters. To make the material more interesting, we have drawn on current and historical examples. These examples should be taken in the context of the purpose of the book and are not meant as complete history.

Who should read this book? The traditional answer is, "Anyone who works with projects." Today, the definition of a project has been expanded to include recurring situations, one-time crises, and dealing with difficult issues, so that project management applies to more people. The right answer today is, "People who have to deal with complexity." What do you need to know to get the most out of the book? You do not need a mathematics or technical background. You will be faced with difficult and complex issues—at work and at home—and you need only to have a desire for success based on an organized approach.

Bennet P. Lientz
Kathryn P. Rea

I MANAGEMENT CHALLENGE

1 Projects and Trends in the 1990s and the Twenty-First Century

PROJECTS AND TIME

A project can be thought of as the allocation of resources directed toward a specific objective following a planned, organized approach. Almost all companies and agencies carry out major work as projects. Projects are shaped by their environment: society, time, politics, regulation, technology.

Successful project management and projects are not new. Throughout history, vast projects have been undertaken across generations, many with great success. In this book, we will consider some of these—the pyramids, the Roman Coliseum, the Taj Mahal, and others. People undertaking projects 2000 years ago may not have had the technology we have today, but they often had political and economic stability and a society that took a long-range view of life and the world. Understanding how they accomplished these projects provides insight for us today. We can take these lessons from the past and apply these to current examples.

Over the past fifty years technological events have accelerated. Pressure is on for more rapid results. People seek cause-and-effect relationships within weeks or months as opposed to years and even decades. Immediate gratification and success are sometimes the name of the game. The pace has quickened in the 1990s. People in every era or decade view their period as unique and, over the last hundred years, superior to previous periods. Current times represent a period of geopolitical and social change. Even weather patterns have changed, impacting societies, economies, and governments. The range of political and social issues is wider. Many issues previously given lower priority or simply ignored have now surfaced (e.g., political divisions, problems with large corporations). Technology has rapidly improved in computers and communications. All these factors have impelled us to undertake many new ambitious projects in restructuring, setting up whole new industries, and creating and modifying products.

These changes have impacted all aspects of projects. Time restraints have forced people in many cases to undertake projects with greater payoff in the short term and to neglect the longer term. Patience is often in short supply on projects. Instead of seeing a project through to ultimate completion, many organizations plunge into the project and attempt (remotely or directly) to micromanage it for quick results.

In this chapter, the nature of current and future projects will be examined from the perspective of how projects can be successfully undertaken and completed in this environment.

THE ENVIRONMENT OF A PROJECT

A project is set in time. It is also set in the context of organization, a legal system, a political system, a technology structure, an economic system, and a social system. How do these environmental factors affect projects? How should a project manager and project team respond? Outside factors impact the project, and the project must respond to the challenge. It is a variation of Toynbee's theory of challenge and response. Take a simple example: for instance, the technology of electronic mail. If a company employs electronic mail widely, then its use is accepted and becomes a de facto rule. It follows that electronic mail will find use in the project. But to be effective its use must be organized. This is just one simple example of how technology use in a company shapes a project.

Why is the environment of a project important even though many of the general principles of project management are timeless and apply to projects of all sizes and types across time? The resources, budget, methods, and tools of the project depend on the environment. If today's projects are managed with tools of the 1960s, they are more likely to fail. This does not imply that past experiences do not apply today. Management expectations, budgets, and competitive pressure may be different; however, many principles remain consistent over time. Only methods, techniques, and tools change. We will consider some of the trends in the decade. The challenge is how to combine the technology of the future with lessons from the past to successfully implement projects.

TREND: GLOBAL COMPETITION

As the world economy changes, there is more competition. Improved communications and transportation coupled with freer trade allow for firms halfway around the world to compete. For example, flower growers in South America can grow and ship fresh flowers into the United States cheaper than American firms can do it locally. Other examples abound in a variety of industries. More economies are expanding regionally or globally.

This trend has a number of impacts on projects and the project management process. Projects in organizations tend to be global, or at least regional. This compounds complexity within the project as well as management complexity. More complex issues arise from language and cultural differences. Also, because of increased competition and risk, projects have to be carried out on a more formal basis. People cannot afford to improvise with costly resources.

Consider the efforts by the major automobile manufacturers to build world cars. A world car is one that can, with few modifications, be manufactured and sold anywhere in the world. This is very difficult to accomplish. It requires an investment of billions of dollars and has high risk. All parts of the project must be carefully coordinated. Again, the use of project management is required.

TREND: RAPID TECHNOLOGICAL CHANGE

Change and advances in computer and communications technology are well known. Growth in microcomputer power, data communications, and image and graphics are just three examples. What is unique now is that changes are occurring simultaneously in different areas causing a multiplier effect and "cross-pollination" between technologies—synergy in action.

Taking advantage of technology for competitive advantage is a challenge because it is continuously improving. Where once a firm could implement a computer system for inventory and then just maintain it for ten or fifteen years, now the inventory system will be changed with new technology and will be integrated with ordering, pricing, shipping accounting, and other systems. Technology advances make possible corporate downsizing, business reengineering, and interorganizational systems.

To illustrate how much technology has changed, consider an old computer project of implementing a payroll system in a factory. In the 1960s the system would have had some terminals for entering data, but the basic system was programmed and set up for production on a large mainframe computer. Scores of programmers were required, even though the project was relatively simple, because it was confined in scope.

Today, it is vastly different. The payroll system should link to timekeeping monitoring systems in the factory. People can log what task they are doing using radio or cellular equipment. The system is tied automatically into accounting and benefits. It is an integrated, distributed system. It might link to several locations around the world. Now the system must handle different currencies, time zones, and work rules. As you can see, more opportunity and benefit, but also more complexity and greater risk. Hence, for successfully implementing new technology, we rely on the organization and structure of project management.

TREND: PRODUCT OBSOLESCENCE

The time between the initial design of a product and the production and sales of the actual, finished product used to be measured in years. Today, we have teams of people working in projects designing, prototyping, developing marketing strategies, and setting up manufacturing and distribution on a compressed time schedule. With computer-aided tools such as computer-aided design and manufacturing, the actual work processes are more integrated and abbreviated. Projects require more coordination and have to be more exact.

The stage is set for increasing pressure to produce new products more quickly and less expensively. There is also pressure to come out with new versions of the product. The many varieties of the Sony Walkman radio, TV, CD player, and camcorders attest to this. Developing and supporting products must be more organized and integrated. This is such an important area that we will devote Chapter 10 to product management and its relationship to project management.

TREND: ORGANIZATION DOWNSIZING

Organizations are under pressure to improve financial results and to reduce their head count for greater efficiency. The downsizing relates to business reengineering and using new technology. However, in many cases, departments just downsize, and the remaining people have to pick up the pieces. Downsizing creates projects in developing plans for downsizing and sorting out reorganization.

Downsizing also has an impact on project teams. Teams are smaller. There is less administrative support. People on the team and the project manager must do more with less. This tends to force people to use technology in the hope of increasing speed and effectiveness as well as augmenting scarce resources.

Projects also face more technical challenges. With downsizing many of the senior people who had knowledge of systems, processes, and the organization have left or retired. This void is felt in the projects as people struggle to uncover how things work and why they were designed the way they were. Twenty years ago, one senior person on a project team was worth his or her weight in gold because of the value of his or her experience and knowledge. This experience and knowledge eliminated learning curves and cut out work. When these people are no longer available, the project team may make the same mistakes that were made over 20 years ago.

TREND: BUSINESS REENGINEERING

Business reengineering is the process where an organization analyzes its basic business processes (product, manufacturing, sales, accounting, etc.) and may totally restructure them across departments with greater automation and management control. Business reengineering can result in major savings. In one company, an accounting staff of over 300 was reduced to ten through reengineering. In another we were able to reduce a banking staff by 80% while increasing efficiency by 100% with the same workload. Change is dramatic because we are willing to discard the old systems and procedures along with the old organization.

To undertake successful reengineering requires a major project effort crossing many departments, which heretofore, had little contact: a challenge for the project manager and project management. The end products or milestones of business reengineering can be vast in scope with new business processes, a new organization, new job descriptions and duties, different computer systems, and new control systems. In a reengineered organization, there is greater accountability, measurement, and control—all of which have to be established in an organized manner.

TREND: EMPOWERMENT

More companies are experimenting with empowerment of employees. The organization gets flatter and leaner. The people who are left in the organization often become empowered and accountable for improving their own business processes.

This empowerment can lead to chaos if every department adopts a different approach. A standard that is often imposed is a project management structure. A similar project structure across the organization can lead to greater synergies between departments and increase benefits across the company and the entire organization.

TREND: FOCUS ON QUALITY AND CONTINUOUS IMPROVEMENT

The Japanese pushed quality, and now everyone is talking about and trying to improve quality. Companies have found that by improving business processes and quality, sales and profits increase. Quality cannot be implemented by just telling people to work better and smarter. There must be a management philosophy. The management strategy has to be backed up by training projects, implementation projects, and quality control projects. Many firms now have had quality project teams in place for years. They stay in place and continuously improve. Quality can be traded off with cost and price to obtain value. The question is, "What level of quality is cost-effective?"

In decades past if you carried out improvements in an area, you might leave the area alone when you finished. You would let it stabilize and then measure the results. This is a luxury that we cannot afford today. Competition is biting at our heels. We have to continuously improve and keep looking for more savings and increased performance.

TREND: MEASUREMENT

Many organizations began measuring their basic business processes, such as assembly, inventory, marketing, and sales, through industrial engineering over 60 years ago. Today, integrated computer and communications systems allow for the measurement and collection of vast amounts of data. We can implement measurement on a constant basis. Continuous measurement feeds into quality, business reengineering, and continuous improvement.

A measurement program is sometimes seen as a set of projects that are carried out in parallel and are performed on a recurring basis. The notion of program management and its ties to project management are a subject of Chapter 9.

TREND: INTERORGANIZATIONAL SYSTEMS

Going hand in hand with some of the above trends is the growing interdependence among suppliers and customers, competitors, and generally within industry segments. Many industries now have standards for interchanging data electronically. Invoices, work orders, shipping information, and even project information are transmitted in standardized electronic formats. EDI (Electronic Data Interchange), has helped to change the industries and company organization structures by making consistent information available faster. Implementing and supporting this and other interorganizational systems requires the organization of a project and project management.

For example, many automobile manufacturers have established ongoing project teams to work with suppliers to move to EDI. By moving to EDI, the companies can get closer to just-in-time (JIT) for manufacturing. The Department of Defense mandates EDI with its contractors. Retail firms such as Wal Mart and K-Mart have successfully implemented EDI and have developed different relationships with their suppliers. These projects are larger and more complex than projects of twenty years ago, which tended to exist in one company or division.

The Internet and World Wide Web have made possible electronic commerce wherein structured electronic financial transactions are available to many individuals and small companies.

TRENDS IN PROJECT MANAGEMENT

Project management itself has been affected by technology. Traditionally, project management and planning were performed by professional schedulers or managers. Project management focused on a single person. This has changed. The modern tools are network based and support information sharing and a team approach to problem solving. More often, we would not consider doing project management today without electronic and voice mail, the Internet, the Web, database management systems, and intranets. These technologies put the team in touch with each other and support a group effort.

TRENDS ARE BLENDING AND COMBINING

Taken together a company can be reformed and totally changed. Suppose a manufacturer competes in the world market and feels pressure to improve. The company decides to actively move into the world market (globalization). In order to succeed, its products must be competitive. It must become more efficient. It undertakes relations with its suppliers (interorganizational systems) and business reengineering. It deploys new technology to support this. Products are improved through integrated design and development projects. Quality is improved in the process. People are empowered to improve their processes in an organized way. The example is summarized in Figure 1.1.

ORGANIZATION EFFECTS

Consider these trends together or even leave some out and what is discovered? The traditional hierarchical organization is under question. In some organizations it loses stability and becomes something that can be changed if there is enough benefit. In mathematical terms, the organization

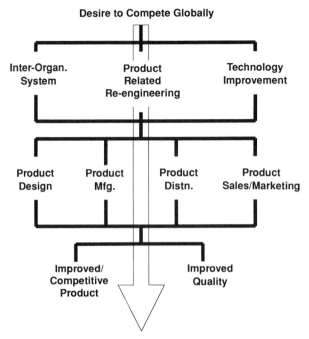

FIGURE 1.1   Simple project in product global competition.

is a variable that can be altered rather than a constant to be lived with and endured.

This organization change gives new importance to the project organization. In all of the trends we have discussed, the project team and project manager are the agents of change and improvement. In some companies the stability has been transferred from the hierarchy to the project organization. But this is not true in every organization. Most companies still have traditional organization structures. A major concern and subject of this book is how projects and project managers address the traditional organization. Their approach can spell success or failure of not just the project, but of the entire organization. Look at the mixed results of the massive investment of companies such as General Motors in automation at the factory level. The effect of only a partially successful project had a profound impact on management, employees, and shareholders.

THE IMPACT OF TRENDS ON PROJECTS

First, projects tend to assume more importance to management because the projects are touching on the lifeblood and arteries of the organization.

Because they are more important and have more visibility, there is an increased need to do the projects right and achieve success.

A second impact is that the projects tend to be complex. They do not necessarily get larger. So many organizations have been burned by the failure of large projects that many firms discourage large projects due to risk.

Transformation of project management itself is a third impact. With the technology available, we are more capable of managing projects that are global, cross-industry, large scale, and integrated across the company.

Pressure to achieve results quickly also impacts projects today. The company may have started the projects of change too late. Management is under tremendous pressure to show results and fast. Thus, not only is the project visible, but it can be under severe time pressure.

Resources are more limited today. With fewer people and less support, we are all doing more for ourselves. The project team has to be more self-sufficient. This links to empowerment and accountability. It also leads to greater conflict and competition among projects.

Another impact is flexibility. When we embark on a project today and are able to demonstrate some results, we may have management change the scope rapidly to expand the benefit and impact. People are not inclined to wait if something is proven and works. In some cases, this occurs prematurely before benefits and impacts are known.

There are three areas where we can look for support to deal with these issues and impacts. First, we can learn from successes of the past in terms of how to organize and carry out projects. Second, we can use the methods that have been employed on thousands of projects over the past 100 years. Third, we can turn to technology for tools to support the methods.

LESSONS OF THE PAST

Massive movements, political structures, and other human achievements have been the result of successful projects. These endeavors were undertaken under limited circumstances and with few formal tools and methods. We have written records and can make inferences based on history, archaeology, and other fields.

When we examine these examples, we can discover a number of lessons in different areas:

- How a project team was assembled and coordinated
- How the projects were managed and organized
- What methods were used to carry out the projects
- How changes were controlled and introduced

METHODS OF THE MORE RECENT PAST

There is a wealth of more recent successful and failed projects on which we can draw knowledge. The use of project management charts GANTT and Program Evaluation and Review Technique (PERT) in nuclear submarine projects, large product development projects, and public and social programs are some examples.

These methods are useful if they can be placed in the context of today's trends, impacts, and technology. If we merely forge ahead and use them as they were used thirty years ago, we will meet with, at most, limited success.

WHAT IS A PROJECT?

We have already defined a project as the allocation of resources directed toward a specific objective following a planned, organized approach. Figure 1.2 shows the components or parts of a project. At the top is a project objective. This is the overall purpose of the project. But objectives are vague. So we need a strategy to support the objective. The strategy identifies how we will achieve the objective of the project.

The basic project document is the project plan. The project plan lives and breathes and changes as the project progresses or fails. It evolves. The project plan must define the work to be done, the resources required, the methods to be followed in the project and its management and control, the tools to be used in support of the methods, and the schedule of work. We also include the milestones or end products that the work will produce. The project plan contains the approved budget.

FIGURE 1.2   Components of project.

The structure of the project plan is supported by the resources of the project. This includes the project manager and the entire project team who will participate in a full- or part-time role. Advisors, consultants, and others who play cameo roles in our project drama are also members of the team.

THE PROJECT ENVIRONMENT

In the 1960s through the 1980s we would have stopped with Figure 1.2. Not so today. We have to be concerned about the factors external to the project that impact the project. Table 1.1 shows some of the external and internal factors affecting the project. These factors cannot be ignored.

WHAT IS BEING MANAGED?

This seems like a simple question at first. The project resources are managed. This includes the budget, the people, and other resources. People often come to the project from different organizations. Each organization

TABLE 1.1   Examples of Factors Impacting Projects

A. Technology
    1. New tools to make products
    2. Technology to test quality
    3. Improved technology to sell and distribute products
B. Competition
    1. Ideas or concepts taken from other industries by competitive firm
    2. Competition improves services and products
    3. Competition invades market intended for your project
C. Government regulation
    1. Increased government reporting
    2. Regulations impacting your subcontractors
    3. Certain tools in the project are regulated or prohibited
D. Politics
    1. New party in power—changes in priorities
    2. Project cancellation or redirection
    3. Project acceleration
E. Cross-impact examples
    1. Technology that makes different products more competitive with yours (electricity vs. natural gas)
    2. Government regulation restrains your competition—reducing the need for your project

has its own interests, agenda, and management contributing to the coordination workload.

Methods and tools are part of project management. Why can't we just identify and specify a set of methods and tools? Success will be partially based on what is selected and how it is employed. Today, tools and methods are interrelated. If we don't use one tool right, it may prevent the successful later use of other tools.

Methods are the ways in which we do the work in the project and the ways we control the project. A good method requires not only guidelines, but also tools that help ensure that the method is carried out. A good method for a car is to rotate the tires on a periodic basis. Doing this by hand without the proper tools can be not only time consuming, but could damage the car. So we need tools to jack up the car and remove the tires.

A tool does not exist in a vacuum. Its use has to be learned. Learning is not enough. You must become skilled and proficient in its use. You need not only guidelines, but also some expert to talk to when questions arise. Tools support methods, not vice versa.

Beyond the project resources, the project manager must deal successfully with problems and opportunities that arise in the course of the project. We call these issues. We want to give you suggestions on managing issues. We want to explore how to turn a negative issue into something that has a positive effect on the project. Issues that are not addressed increase project risk.

PROJECT RISK

Much has been written about project risk. People have tried for years to quantitatively define and measure it. Techniques for managing project risk go back to ancient times. What is different today? There are more potential areas of risk. Impacts of risk can be more dramatic.

External factors influencing risk are beyond our control. How can we minimize risk from something over which we have no control or effect? That is a reason why we give attention to tracking and measuring risk and to establishing contingency plans.

Risk is the product of the likelihood of failure and exposure. Exposure is the potential cost in money, resources, and schedule. With effective project management we can reduce both the probability of loss and exposure.

PROJECT CHANGE

We not only have to get a project started right; we also have to manage the project as work is done. This means more than tracking and control. It means the control of change in the project. We devote time to project change: how to identify when changes are needed; how to introduce change; how to measure the impact of change.

WHAT MAKES FOR SUCCESS AND FAILURE?

We all want our projects to be successes. Who wouldn't wish this? But it is not that simple. Today, we tend to implement more quickly and see short-term results faster. But we don't know about the long term. The time horizon must be defined. When nuclear power was first developed, it was viewed as a great success and savior for energy, medicine, and a variety of social problems. Although it has had many benefits, the view today is mixed. In short, something that appears as a success in one year may be a failure two years later.

Perspective also affects whether we see success or failure. If we are in poor country and the standard of living improves, we see success. On the other hand, from the perspective of a more developed country, the situation may look life failure.

A project can achieve its objective and yet fail due to side effects. Take a large dam in Africa. It holds water to control flooding. It can generate power. But it may silt up. The cost of the power infrastructure may be more than that of imported oil. The lack of flooding may mean that the country has to import vast quantities of fertilizer. The side effects may outweigh the engineering success.

What do we do? Consider and measure side effects. The completion of rapid transit systems may change population living patterns and create overcrowding and sprawl. These were certainly not intended by the designers, but are predictable. The effects of the project end products and milestones as well as the objective need to be considered.

There are other impacts of a project. A project might be viewed as a failure, but it showed that a specific method or tool worked. This is true of early work with arches and other architectural structures in medieval Europe. The first ones were crude and may have collapsed. But people learned from the failure and built the next ones better. This is true today with various anticrime federal test programs. As we proceed, we will be learning how to address projects in today's environment with modern tools and methods, while we attempt to learn from successes and failures of the past.

EXERCISES

In each chapter we will be suggesting some things to do that relate to what we discussed. These are, for the most part, thinking exercises and are important for implementing the ideas in this book. That is what we want you to do. This is not a book to sit on a shelf or table. An art book would be a better choice if that's what you're looking for.

We have discussed the framework for a project as well as some of the factors that affect a project. This has been done in the context of some of the current trends in society today. You should be able to take a project mentioned in the news and define the impact of the trends on the project. You can also attempt to define the components of the project. Why do this? You need to develop the capability to think of systems and work in the context of a project.

SUMMARY

This book provides approaches for managing today and tomorrow's projects with modern methods and tools tempered by the lessons of the past. A road map for the book has been presented. We will be examining organization relationships with projects in the next chapter. Some examples will be considered in Chapter 3. With this background, we are ready to plunge into project management. We first consider the overall approach and then turn our attention to issues, methods, and tools. Beginning in Chapter 4, we will discuss a modern, typical project at the end of the chapter.

Our approach is to take a very pragmatic and not theoretical view of project management. As project managers ourselves, we emphasize the realities of the world of project management, including politics, conflict, and the exhilaration of success.

2 Projects and Project Management

ARE PROJECTS DIFFERENT?

Why can't projects be done as ordinary work in regular departments? Why have a separate structure? Very small projects involving one or two people for a short time can be done in a department. It is a senseless argument that anything that is a project has to be done in a separate organization. However, with larger projects there are many benefits to the separate organization approach. Where to draw the line between large and small depends on organizations and the nature of projects. With rightsizing and accountability more work is being performed as projects.

THE TRADITIONAL ORGANIZATION

We will refer here to the normal organization as line or functional. Staff departments are part of the normal organization and can be viewed as functional or line. Although many have attacked current hierarchical line organizations as bureaucratic and unable to accomplish goals efficiently, the traditional structure has lasted a long time—for several reasons. First, the line or functional organization represents stability for employees. It is a home. There are existing relationships and controls within departments.

A second reason relates to communications. A standard organization has an existing vertical communications system already in place. Communications channels are established and tested. To see how important this is, consider what happens when police from different cities try to coordinate a joint project or see how long it can take the United Nations to get started on a peacekeeping mission. A department does not require time to mobilize. If we start a project, we have to set up an organization structure and allow time for them to get started. Quicker reaction is possible.

In a department there are existing budgeting and cost accounting controls in place. If you want to get something done, you could just transfer money and support to and within the organization. It is easy and proven.

There is better technical control in line organizations. People who have hard-to-find skills can be used on many different activities. If they were dedicated to one project, they would not be available to other projects. Related to this is flexibility in the use of people with respect to tasks.

Line organizations are set up to handle routine and recurring tasks. Mass production is an example. No one would consider having a large project operating a factor forever. The project might set up the production line and tune it. After this period, management would then transfer to a standard part of the organization.

For these reasons, many companies have attempted to undertake projects within current organizations. This has been true from ancient times to the present. It is the default or the path of least resistance. It will continue to be true in the future because many general managers are more comfortable with their standard line organizations. Because some people do not seek to learn from the past, they repeat the failures and errors of the past.

FAILURE OF LINE ORGANIZATIONS IN PROJECTS

Suppose that a project is to be started. The project has management support. In a meeting on the project, almost as an afterthought, someone asks where the home of the project will be. There is silence. Because they want to go on to a new agenda topic, someone suggests a particular line organization. There is agreement and the line organization is stuck with a new project. Although it does not happen this way all of the time, it is frequent enough to serve as an example. Here are some of the problems.

Lack of Focus and Attention

The department already has a list of tasks to which it is being held accountable. It will likely be responsible for these normal tasks ten years from now. Off on the side or added to the list is this project. The project is a onetime thing. It will vanish when it is completed. The project also differs from the normal work and is unfamiliar to the department staff. Given these factors it is not surprising that many departments then pay little attention to the project. The project may get treated as another task to be managed and reported upon like everything else.

Inability to Cope with Different Project Characteristics

A project is different from usual work. It has a single focus. It requires almost constant attention. It is a child in the midst of a family of adolescent and adult tasks. Some managers cannot understand the differences, which include (1) new relationships with other departments; (2) tighter time and budget pressure; (3) use of different methods and tools; (4) different reporting structure to management.

Feelings of Being Used and Exploited

People in the line department may feel used and abused because they are now being called in to work on this project in addition to their current work. If they were assigned to work on a project in another department, that would be one thing. But it is here in their midst. Employees now feel split between the project and their normal work. Guess which gets priority. Some employees will look for any excuse not to work on the project. With mixed feelings and commitment, it is no wonder that the quality of work on the project is, at best, mixed.

Lack of Project Experience

Management may lack experience in coping with projects. So what? They are experienced managers and can run projects just as well. Not necessarily true. Issues, problems, and opportunities were in fact visible very early in the project, but were invisible because management lacked project management skills and experience.

Example of a Project in a Line Organization

We start with a moderate-sized project that crosses three departments (A, B, and C) and involves six people. The project is assigned to department A because department A will be the most involved. The manager of department A does not want to look bad to management and so accepts the burden of the project. The manager of A does not draw up a project plan because he or she thinks that the project can be run out of the department with existing resources and standard methods. "People will just have to work a little harder."

The manager returns to department A and announces that they have just been given responsibility for the project. Words are said about how important the project is to the company. "Everyone will have to make an extra effort. We want to do a good job and show our department off." Brave words—with little content. There is no project manager and no plan. It sounds to the staff like another set of activities and more work.

After the meeting, the manager realizes that something must be done and casts about for someone to be the project manager. Sometimes, the manager may even attempt to be the project manager her- or himself. Most of the time this does not last long due to other duties and a feeling of being lost in the detail of the project.

Think about an average business department. At any given time who is available to be a project manager? As the manager, are you going to give the job to a proven performer? Are you going to disrupt the department by using one of your best people? Not likely. So the manager casts about for who is available or who can be assigned without disrupting the department. These are obviously the wrong criteria for picking a project manager. Unfortunately, it happens all of the time. You might think that upper management would make the same mistake. Not as true. They would give the project visibility and support and attract a good person from some department. The department would then be told to get along without them.

The new project manager who lacks experience with projects is now named. The project manager is told about the project by the department manager and also given the direction to get on with the work. The new project manager wants to do well and sees this as an opportunity. Well then, better show some results quick. The project manager may attempt to assemble some team members. People in the department may be reluctant to become involved in the project. With lack of management visibility and being buried in the organization, do you think that the project is going to attract good team members? Hardly or very seldom.

Wait, it gets worse. The project manager goes to the other departments involved (remember B and C?). The project manager shows up and indicates that he or she is the project manager and is there to identify people for the project. The managers of B and C are not stupid. They know the organization and see that the project manager lacks experience and does not have an overall plan. Will they assign their better people to the project? Unlikely. They also may think that if management perceived the project as important, they would have set the project up differently. With no visible major management commitment and with a junior project manager based in a department with which you may or may not get along, what would you do as the manager of department B?

Let's next assume that the project commences. Project reporting will be done through department A. The department A manager will probably report on the project in addition to department budget, capital expenditures, general staffing actions, and other items. The project is submerged. There is little time to devote to it in a 30-minute update of the department activities.

Suppose that an issue surfaces. Problems and opportunities don't often just appear from thin air. There are symptoms of the issue that surface first. If the project manager is inexperienced and too busy working on the project due to the lack of commitment of the departments, then the symptoms may go unnoticed or be ignored. Even if they are recognized, there is the danger of misinterpretation. The issue may fester and worsen.

Typically, after some visible impact, the issue will be identified. Department A manager may panic or try to solve the issue within the confines of the department. The department looks good because they tried and succeeded in addressing the issue without outside management involvement. The issue by this time is likely to be beyond the scope of the department. A disaster is in the making. As an aside, it does not necessarily follow that if the foregoing chain of events occurs, an acceptable project structure will finally be created. In several examples, senior management still didn't learn from the experience and merely assigned it to department C! Thus, history repeated itself. After the second time, the project was killed.

So If a Separate Project Idea Is So Good, Why Not?

Let's take the other extreme. From the foregoing discussion we will assume that the idea of separating project structures from the normal organization is a good one. This must be tempered by the fact that a company cannot operate for a prolonged period of time on projects alone.

Attila the Hun, while having short-term success, failed in the long term. You need an infrastructure within the organization to succeed in the long term. The dilemma is as follows. How to get benefits from both worlds— completing the projects done while keeping the organization intact. Enter the matrix organization.

MATRIX MANAGEMENT

A matrix organization appears as a table of row and columns: line departments are rows; projects are columns. People in the organization belong to a row or line organization. When they are required on a project, they move to that column. We could create an artificial column called *unassigned* for people who are waiting for an assignment or have just completed the assignment.

Companies that carry out government and private projects have some-times organized themselves into matrix organizations. Results have been mixed. There is also a middle ground—an organization with a few projects and with standard production or operations processes.

Table 2.1 illustrates the advantages and disadvantages of the matrix organization. To understand these, consider how the organization func-tions. Management identifies a project manager. This creates a column. The manager now goes to the various line managers for resources. The line managers have to cooperate because the organization focuses on projects. People are committed to the project (they move to the column of the project). The project team is now assembled.

When the team member's work is done on the project, the project manager returns the person to the line department (the person moves from column x of the project). If this sound mechanistic, that is because it is mechanistic in some organizations.

The project manager wants to bring in the project at the lowest cost. This encourages the project manager to release the person as soon as possible. The line manager sees the people who are not working on projects;

TABLE 2.1   Advantages and Disadvantages of
Matrix Organizations

Advantages
1. Very good for project-oriented companies
2. Ensures that people on projects are utilized (as otherwise they are returned to the pool)
3. Project manager tends to be powerful in getting resources
4. Accountability and tracking of projects improved
5. Possibility that people who move between projects can build skills
6. Provides formal structure for projects of medium to large size
7. Ability to track what people are working on in projects

Disadvantages
1. Good people will be in heavy demand for projects; others, who are not so good, will sit in the unassigned pool.
2. Difficult to assign control between project and line management.
3. Line managers tend to be weak.
4. Projects with long lives tend to be confused with line organizations.
5. Difficult to share resources between projects
6. More difficult to have lessons and skills cross projects—less chance for organization history
7. Project prospers and traditional organization suffers
8. More difficult to anticipate resource needs and to staff for requirements
9. More difficult to address small projects

they are overhead. They can do research, be trained, and do administrative work, but they are still overhead. The line manager is under pressure to keep the overhead down. The manager works at assigning people out to projects. Good managers are identified sometimes by their overhead or success in marketing their staff. Some accounting firms operate this way. In the perfect world we would realize all of the advantages of Table 2.1. The project manager has authority over the project. Functional or line managers know that they have to serve the projects. Responsibility is shared between the project and line managers. Key people can be shared among projects. A strong technical base can then be established. There is shared experience between the staff in the line department, and conflicts would seem to be avoided.

The world is not perfect. Supporting this structure means greater management and administrative costs. What happens to employees who are not often selected for projects and who have other skills? Unfortunately, they may not last long. The environment can be a pressure cooker. The project managers try to find the next project to avoid being returned. There is a constant tug of war among projects and line organizations. A large project that goes on for some time can almost swallow up line organizations. Stability is jeopardized because of the demand for projects. Focus is on short-term goals of projects and not the long-term objectives of the organization. A matrix organization best fits special situations, such as consulting, planning, accounting, or other similar firms that are predominantly project oriented.

THE MIDDLE GROUND

In between these two relative extremes, there is a middle ground. Small projects are kept within the department, while larger projects are designated as separate entities from the organization for the duration of the project. Later, we will consider how the transition occurs when a small project grows and what happens when a larger project winds down.

In order to visualize how this arrangement operates, we have to consider the differences between the project view and the line or functional view of the world. There is a natural and fairly constant tension between these two. Taking the view of the line organization, project managers often feel that the line organization serves the projects, whereas line managers feel that they should have control because their resources are being used in the project. Both views are valid.

From a project view, the project is a joint venture within the organization. Organizationally speaking, it is multilateral. To the line organization, the project is still part of the organization. The view is unilateral. The project view is short term and finite. The functional organization sees itself as lasting as long as the overall organization. The functional organization resembles a pyramid, although today people are pushing for shorter and flatter pyramids. Projects are managed on a peer-to-peer basis, reinforcing management thinking geared toward smaller, more efficient organizations.

SETTING THE STAGE FOR SUCCESSFUL PROJECTS

Enough about the background and playing field of the organization, let's turn to how the arena can be established for successful projects within standard businesses and agencies. General guidelines will be presented that apply to all projects. In later chapters we will discuss how to carry out a single project with success.

GUIDELINE: RECOGNIZING PROJECTS

Management must recognize that projects are an increasingly important part of life. There should be a regular, formal process for setting up, overseeing, directing, and terminating or redirecting projects. This can be accomplished by recognizing the inherent limitations of any standard organization. The project is not an apology or excuse for the failure of the normal organization, but is a fact of life to get some specific, important piece of work done. Managing projects in an ad hoc manner is a prescription for later problems. If the projects are organized on a consistent basis, reporting and control can be more uniform across projects.

With more recognition of projects, people in the line or staff organizations can see what is expected of them. There is less uncertainty when a project is established. Employees know that they may be selected for a project and what this means. If it is part of training and orientation when an employee joins the organization, this makes the process even easier.

GUIDELINE: HAVING A PROJECT TIER STRUCTURE

Projects have different characteristics depending on their objective, scope, and other factors. There are differences in risks, technology, time pressure, and so forth. To reflect these differences, a tier or set of project levels can be established. This system does not have to be complex. Otherwise, people will spend time arguing over to which level a project belongs. Four or five levels are usually sufficient. At one end is a simple project of one person or several people within a department. At the other end is a very large project that is established totally outside of the current organization. The project levels in the middle can be defined based on the characteristics of the firm or agency. This approach allows for flexibility and the capability to use different methods with different levels. It allows for different levels of visibility with management. Additional details will be provided later.

PROBLEM: SCARCE RESOURCES

We often face a dilemma. Many organizations depend upon a few key people with critical knowledge and/or skills. If these individuals are assigned to projects, they are lost to the line organization. What should we do?

This problem will persist and worsen in the twenty-first century. A solution appears to be restrictions on the assignment of staff to projects. Here are some guidelines we have implemented:

- Critical resources must be managed across multiple projects above the level of the individual project.
- Resources cannot be assigned full time to one project for an indefinite, extended period. They must be released and time shared among projects.

We have to create new projects for specific people and other scarce resources by extracting all of the tasks that apply to them across multiple projects. In parallel, we seek to gather their knowledge and lessons learned, and apply these to reduce dependencies in the future. Examples of critical resources are: an engineer who has unique design skills, a programmer who is the unique person who knows about specific software, and expensive equipment required by several projects.

GUIDELINE: PROJECT TRANSITION PROCESS

If we have levels, then we have to define an orderly transition process by which a project is moved up or down between levels. Without an orderly process, changes will occur and the controls, methods, and tools will not be appropriate to the expanded or contracted project.

What does a transition process include? First, we need a means to make the decision by management to change levels or tiers. That is, we need an organized process whereby the project manager can approach management for a change. Next, we have to supply some guidance and examples as to how the process works. This will give the organization and the project managers some ideas on how to proceed and what to look for in terms of the need for change. Here we want the project manager to be able to approach management and not try to resolve certain issues within the project. Otherwise, the issue will not surface to management soon enough.

After a decision is made on changing levels or tiers for a project, there should be a series of formal steps to implement the change. The following changes may be needed:

- The project plan may have to be updated and restructured.
- Additional resources or fewer resources may be needed by the project.
- The method of controlling the project may be changed.

These and other changes should be implemented as soon as possible after the decision of change as been made. If there is too much of a time gap, then the old project structure and plan continues while the project changes and so the project is misaligned. Some organizations require the project manager to report back to management on the new, changed project.

GUIDELINE: PROJECT MANAGEMENT METHODS AND TOOLS

Imposing the same methods and tools on all projects, regardless of size or risk, is not feasible. However, a set of approved methods and tools can be created and supported. Methods include how to set up a project, how to report on a project, how to address risk and issues in the project, and other activities. Tools include project reporting software and forms, project management software, charts and graphs, and other aids in project management.

The support comes in the form of the following:

- Management endorsement. Management has to openly endorse the methods and tools and not equivocate under pressure.
- Training. Training has to be supplied to teach the methods and tools in the context of the project.
- Expert help. A center of expertise has to be established by one person or a group. There has to be someone to go to in case of questions.
- Control and validation. Management has to establish a method for ensuring that the methods and tools are in use.

Not all methods and tools apply to all projects. For example, take the earlier guideline on levels of projects and define a set of methods and tools for each level. The more complex and the larger the project, the more methods or tools might be used.

GUIDELINE: DEALING WITH LARGE AND SMALL PROJECTS

We have identified setting up tiers of projects as a guideline. Small and large projects are at the extremes. Many firms do not give enough attention to these extremes. On the one hand, small projects involves less effort and, probably, less risk. On the other hand, large projects may be regarded as very rare events. Even in a nation's history large projects do not come along every day. Note, however, that for the past fifty years, there have been more major projects. Why is this occurring? We have already pointed to complexity. Another reason is that organizations gain confidence that they can carry out the projects successfully. There is confidence that almost anything is possible (lunar landing, ending poverty). Technological advances and new technology products help to reinforce this belief.

Structure is required for large projects to ensure consistency for all projects. Structure will also help people in smaller projects who have never carried out a project in the past. Another reason for structure is to support the transition process between levels or tiers. A small project should have at least minimum guidelines; a very large project should have guidelines on how to break it up into subprojects.

GUIDELINE: PROGRAM MANAGEMENT STRUCTURE

Programs are recurring projects. Preparing an annual report is an example of a program. Because each occurrence of the program is a project, we can reasonably expect to have guidelines for the project. We also need guidelines for programs overall. We want to learn from past projects and improve future projects done within the program. What could these guidelines look like?

- Longitudinal analysis. There could be a requirement to analyze the projects and refine the project guidelines.
- Program process. There could be development of general steps to carry out the program.

The critical comment here is that projects must be established and managed in parallel to line organization. Given the number of programs that even moderate-sized companies undertake, the structure of guidelines is important and tends to serve the firm well.

Example: Island Warfare in the South Pacific in World War II

When the United States carried out the early landings in the South Pacific in 1942, amphibious warfare was quite new. It had been done before in World War I, but the weapons and situation were different in 1942. The landings at Guadalcanal were a powerful lesson to the American military. Who went ashore first? How were enemy positions "softened up" prior to the landing? How did the vessels get inside the coral reefs? Which elements of logistics support were landed first? These and many other issues were addressed through reviewing the first landings. The work of designing and configuring amphibious warfare was broken out as a separate project.

By 1944 and even as early as 1943, the process had been refined. The program of amphibious landings was well established. But the learning process continued into Korea with the Inchon landing in Korea behind enemy lines. The Inchon landing was carried out at one of the worse landing places to achieve surprise. It would not have been possible without the learning experience of the Second World War.

Example: Aerospace Projects

The concept of levels of projects has been widely adopted in the aerospace and defense industries. When a new weapons system is started, a design project is the first form of the project. If the design is successful, a prototype of the system may be commissioned. The project now expands and changes from planning and designing to engineering. The project manager may change along with the project team, scope, and other factors. If the prototype is successful, then full production may be undertaken. There is again another project shift.

Example: Consumer Products

A firm may decide to investigate making a new type of soap, radio, or some other consumer good. A product manager may be appointed for the new product. The product will be specified and designed. A marketing plan and packaging strategy will be developed. A production approach will be defined. After this planning, management may choose to develop and test the product in a small market.

Testing in a limited market is a substantial focus of marketing books and courses. It involves sampling, analysis, and other activities to evaluate pricing, packaging, impact on other company products of the same type, and impact on competitors. If the product results are less than successful, the product may be dropped or moved into a caretaker status (moving the product down a level) for potential later consideration.

If the testing succeeds, then the decision might be to move to a national market. The project level changes. Later still, it may be decided to test the product in certain international markets. Although the project progressed through various stages, the product manager may have remained the same. The product approach gives consistency across the various different projects.

EXERCISES

In your organization, think about some of the projects that have either gone on in the past or are active now. Analyze them in terms of whether there are management guidelines for the projects, or whether each project goes it alone. How large a project could you have without having to follow guidelines and rules, and use a defined set of methods and tools? Another exercise is to group similar projects and determine if there is a logical way to break down the projects into categories based on size, number of departments, duration, and other parameters of a project.

SUMMARY

We have learned that we cannot avoid projects. We also cannot ignore the existing organization. The trick is to perform the projects while still keeping the current organization and its activities intact. This challenge has been a major area of study and controversy in project management. That is why we have developed various project organization models such as the matrix organization. Although such an organization may suit some project-oriented firms, it does not serve a standard organization well.

We tend to manage small projects within departments and split out larger projects on a case-by-case basis. The compromise is to adopt an organized approach whereby projects can exist within the firm or agency, but have some autonomy from the functional or line organizations. In order to be successful, the technique selected must allow for the strengths of project management and line organization to contribute to the project.

3 Lessons Learned from Projects

In Chapter 1 we defined what a project is and some of the ingredients and factors involved. Then we examined how projects fit or do not fit within the typical organization. Getting information from completed or failed projects provides us with lessons learned. What can we learn from past projects?

PROJECTS: THE GOOD, THE BAD, AND THE UGLY

Projects of the past can teach us about project organizations, methods, and techniques. We should not expect the past to present itself as neatly structured lessons learned. We have to decipher past experience. What do we collect? From whom? How do we interpret the information gathered? Many of these projects were not called projects at all but were nonetheless regarded as different from everyday life.

How do we analyze a project of the past—recent or ancient or current? Here is a step-by-step approach:

1. Identify a situation that is interesting to you because of your background or your current job. This situation does not have to be labeled as a project. It does not have to have advanced technology. It should have interesting issues and results.

2. Gather reading materials on the situation from different viewpoints. For example, for an ancient example we might draw on history, archaeology, and sociology. For a modern example, beyond news reports in magazines, and so forth, we might find reports or books on the subject. Search the Web for additional information.

3. Develop a concept of the structure of the project. That is, determine the key events, milestones, how the project was managed, what methods and tools were used, and what results ensued.

4. Analyze the project in terms of objective, scope, strategy, the relationship between the organization and the project, how the project was managed, issues faced by the project, project team communications and control, the use of technology, the transition after the project was completed, and the degree to which the project was both a success and a failure.

This is the approach we have used in this chapter's examples. Our examples are not intended to be totally complete; also, we recognize that there are different perspectives for some projects. We are trying to motivate the study of project examples from the past and present because it provides insight and is an interesting process. Each example will be reviewed in terms of the project structure discussed in Chapter 1 and the organization issues of Chapter 2. Modern and ancient examples are deliberately mixed.

Our first example is very brief. Napoleon was asked how soldiers should prepare for becoming generals. He urged the repeated reading of stories of the over eighty battles (projects) of Alexander the Great, Hannibal, Caesar, Turenne, Eugene, Frederick the Great, and Gustavus Adolphus.

ALEXANDER THE GREAT AND SUCCESSION: PROJECT MANAGER TRANSITION

As we know from history, Alexander the Great carried out a very successful project of conquest in the eastern Mediterranean, Asia Minor, parts of Africa, and even parts of India. He was younger and in good health during much of this project. The project stopped unexpectedly when he died of disease (probably malaria). There was no succession plan in place, so the empire that he spent so much time and effort to build was summarily and efficiently carved up.

The lesson here is that every important project needs a transition plan. We have to take into consideration the fact that key project team members or the project manager may leave the project quickly. The project still must go forward.

SENNACHERIB, RULER OF NINEVEH: PROJECT MANAGER OF CANALS

In 705 BC Sennacherib became ruler of Nineveh. He first built a protective wall around the city and constructed gardens on the grounds and on buildings. Lions and other animals roamed on his grounds. But with the population growth and the need for water for his gardens, it became imperative that stable water supplies be found in sufficient quantity. Sennacherib built a number of canals. The early ones were not sufficient. So using what he had learned, he directed the construction of an aqueduct and canal system to bring water to Nineveh from 30 miles away. His aqueduct was a marvel of its time. It was a bridge over 90 feet long with arches. Mortar or some other similar material was used to seal the waterway so that there would be no leaks. This would have been an achievement in itself, but he did it in 15 months!

This is another example of the benefit of management support and dedication. It shows what can be accomplished when the priority of the project is set high. The objective and scope were crystal clear.

NECHO: AN IDEAL PROJECT MANAGER

Necho became the ruler of Egypt in 610 BC. He was responsible for two major projects in his life. These projects still stand out 2000 years later. The first was that he managed and directed a project to link the Nile with the ocean. He created a canal that ran east to west. It started at the town of Zagazig in Egypt on the Nile and moved east to about the middle of the present Suez Canal. Boats could then come from the Red Sea to the Nile. He also commissioned an expedition by Phoenicians around Africa. In this remarkable journey, the team sailed from Egypt and circumnavigated Africa from east to west. They entered the Mediterranean and returned to Egypt after two years. During this time, they repaired their boats and actually grew crops to survive.

These two ancient examples show what can be done in projects with management support and determination. It is unfortunate that so few details of the work are known.

ASHURBANIPAL, KING OF BABYLON: THE WOES OF A MANAGER

Ashurbanipal was one of the more successful rulers of Babylon. His armies dominated much of the Middle East. Yet, toward the end of his reign, he had much unhappiness. Although he was much more than a project manager, his words speak to us across the ages of the feelings of project managers who suffer reverses:

> I did well unto god and man, to dead and living. Why have sickness, ill-health, misery, and misfortune befallen me? I cannot (do) away with the strife in my country and the dissentions in my family. Disturbing scandals oppress me always. Misery of mind and flesh bow me down; with cries of woe I bring my days to an end. On the day of the city-god, the day of the festival, I am wretched; death is seizing hold of me and bears me down. (James R. Pritchard, *Ancient Near East Texts* [pp. 1950–1955]. Princeton: Princeton Univ. Press.)

Hopefully, you as a project manager will never feel this depressed. It is sometimes useful to collect and read quotes such as this to see what people have said about their experiences.

ROMAN ARMIES: METHODS AND TOOLS

For hundreds of years the Roman armies dominated the Western world. This did not happen by chance. They were tightly organized into groups or teams. The term then was *legion*. To instill loyalty each legion was given an eagle standard. The Roman soldiers were organized into phalanxes for battle. This was changed later to accommodate new enemies. This showed flexibility in project organization.

Consider the Roman soldier's tools. The Roman soldier carried a large amount of tools, food, and other items. Included were a saw, armor, weapons, a basket, a space, an axe, leather thong, sickle, chain, and several days of food. The Roman soldier was often expected to perform work on roads and do other civilian duties during times of peace. Imagine carrying all of these tools and baggage and then working on a project. What is important is that, with few exceptions, Roman methods supported by these tools prevailed.

A factor contributing to the fall of Rome was that the Roman armies did not improve their tools and methods. The barbarians at the gates did. They used horses more and employed saddles. A saddle is a remarkable tool. It not only allows you to stay on the horse, but also allows you to fight from horseback.

Beyond the interesting historical detail, we can see the importance of reviewing and modernizing methods and tools. We would not want to change the approach in the middle of a battle or project. But between projects, we could assess the effectiveness of what we have used while the experience was fresh and then make improvements.

ROMAN ROAD CONSTRUCTION AND MAINTENANCE: A SUCCESSFUL PROGRAM OVER CENTURIES

Roman roads exist today in many places. They are often still used as is or have been paved over by more temporary asphalt. The roads had a basic purpose—to facilitate the movement of the Roman army. This goal shaped the design of the roads and tunnels. Roads were designed to be straight. This meant that roads were constructed up hills as opposed to going around hills or making cuts in the hill. The grades of some road sections were over 15%—a real struggle for an average traveler.

How did they build the roads? They started by excavating a trench two feet deep or more and checking whether the ground was soft. If so, they then sunk pilings into the soil. The road was built in five (yes, five) layers. Sand and mortar provided the first layer. Then they placed stones tightly together in cement. Next, came a layer of clay and gravel. The fourth layer consisted of rolled sand with concrete. The final top layer was composed of blocks of rock in cement. The materials varied by location and by what was available. Molten lead was used in some places for stability. In some heavily traveled sections, lead stakes were placed in the middle of the road as a dividing double yellow line is today. These roads were designed to last. After all, given the construction techniques and tools used, they should. These roads were really horizontal walls. An average road section needed maintenance only every 80 to 100 years. Compare this with the roads of today.

Tunnels were used to speed up travel. Around Naples, Italy, there were two tunnels. These were 10 feet wide. The height in the tunnel varied from 9 to 70 feet. The greater width allowed for two-way traffic.

Roman roads are more than just an example of a successful project. They are examples of what we call programs. That is, the program of Roman roads was a series of projects that went on for centuries basically unchanged in terms of maintenance. This is an example of such a long program that we could treat it as a continuum.

Roman roads are an example of how long a program and project can work if the basic situation does not change and if the objective, strategy, methods, and tools work well together. The Romans were great builders. We will use this example and others later.

At its height, the Roman empire had several noteworthy circuses for entertainment. The largest, started by Julius Caesar, was the Circus Maximus. It could seat over 250,000 people. The circus consisted of a chariot course over a quarter of a mile in length. People sat in seats on both of the long sides of the course oval. Chariot races were relatively simple to stage and appealed to a huge audience; hence, they were more cost effective than the Coliseum, which could handle more events, but for a smaller audience.

ATTILA THE HUN VERSUS GENGHIS KHAN: THE PROJECT ORGANIZATION

Attila the Hun and Genghis Khan were both conquerers. Many history books and stories treat conquerers alike. But these two were very different.

Both were brutal and effective. But, whereas Attila the Hun conquered and did not leave a legacy, Genghis Khan paved the way for established, ordered rule of one of the greatest civilizations of the world in China. The Mongol rulers carried out a number of innovations in gunpowder, writing, military order, and civilian rule.

This contrast shows what we want from projects. We not only want the project to achieve its goals, but we also want contributions to the body of methods and tools so that future projects will improve and build on the experience of the past. This shows the importance of a postimplementation review in projects.

THE BRITISH MAPPING OF INDIA: MAINTAINING A LONG-TERM PROJECT

In the 1700s Great Britain through the East India Company gradually established its position in India. India became more important to Britain on economic, political, military, and social grounds. The involvement accelerated in the 1800s as Britain came to dominate the subcontinent. At that time, India was not one country, but a collection of protectorates, semiindependent states, and direct possessions. The British really did not know what they had. They needed an organized logistics system to handle tea, opium, and other products. They wanted to protect what they acquired from Russia and other countries. Spanning decades, the survey was one of the most complete ever attempted. Survey towers were erected. Standard measuring tools were used by a team of over 100 people. Starting in the south of India, the project was completed at the base of the Himalayas. Mount Everest is named for one of engineers who was a project manager on the work.

The British project is interesting not because it overcame substantial odds and practical problems, but also because it led to many other projects involving telegraph, railroads, government organization, and the entire ruling structure of India. Without this survey, English rule might have been less organized and enduring. Thus, depending on point of view, British rule of India can be termed a success or failure. Over time you can view it the same way. By considering the project as it continued year after year, then it might seem that the project would never have finished and that it would fail. As each year went by, more infrastructure was added so that there were more benefits that created a greater feeling of success of the project. This provided an impetus to continue the project.

The objective of the project started to be a limited effort in southern India. With its demonstration of feasibility and success in getting information about India, the scope was expanded. The project was flexible enough to accommodate the change in scope. The project manager changed several times; again, the project accommodated the change.

———— EXAMPLES OF METHODS AND TOOLS

Here we will consider a collection of methods and tools that impacted civilization and made possible successful projects similar to the ones we have seen.

EVOLUTION OF A TOOL

When the ancient Egyptians and others first thought about ships, they constructed barges and canoe-like craft. But the Nile made for a unique evolution. The Nile flows from south to north, yet the winds blow from north to south. This meant that with a sail, a vessel could easily move up the Nile. In rivers where the wind and current moved in the same direction, progress was retarded and had to wait for later technology. Similar examples exist for farm implements, tools of war, and manufacturing tools.

This shows that tools change as they are used. Experienced and knowledge combined with thought give rise to improvements. It is the same in project management. Many projects could start with the same tools, but the results might be wildly different. Why? One reason is the different experience and knowledge of the project team. Another reason is the leader or manager who knows how to make best use of the tool.

TOOLS WITHOUT METHODS

We have many examples of tools that were very innovative, but led nowhere. We all know Leonardo da Vinci's diagrams for flight—dreams that were only realized centuries later. Another example was recently unearthed in a museum in the Mediterranean Sea. From the haul or trove collected from the sea was a curious object. A scientist examined it and thought it might contain gears. X rays and other examinations later provided enough information to indicate the structure of the object. It was in fact an early computer! It had differential gears and allowed a person to turn a crank. As the crank was turned, objects turned through the gears indicating the position of the sun, moon, and the known planets at the time—2000 years ago!

A number of scientists have puzzled over some early gold work. A scientist has conjectured that the ingredients were available to make a battery. From artifacts in a museum, the scientist demonstrated the battery and its use in gold plating.

But we are often frustrated when we see tools with great potential, but then the potential lies unrealized. Why? The tool has to be fit within the context of a specific activity. The activity should provide some value or benefit to someone. This activity we can compare with the project. Why aren't some tools used more? Why are people in projects sometimes very conservative? For example, why did the Romans overbuild? The answer is that the people believe that they know what works. As long as it works, why change?

Looking at the examples in this chapter, we can see that to be successful it is sometimes necessary to embrace new tools and modified methods (which stem from the new tools). Today, we are more aware of tools and tend to want take advantage of them if they work for others. In a project there is, however, a caveat. We cannot just pick up and drop tools. Once we start to use the tool, we get committed to it. Dropping a tool and not replacing it raises risks and instability.

EXAMPLES OF PROBLEM PROJECTS

SIDE EFFECTS OF PROJECTS: TRAFFIC CONTROL ON FREEWAYS

Problems can occur when a project is divided into separate teams each with their own objectives. In many places the freeways or turnpikes are managed by one group and the surface streets are managed by another. With freeway crowding the freeway project team considers ways to restrict traffic. They start to look at the on-ramps to the freeways. Ah, they think, if they put tollbooths or meters on the ramps, then the number of vehicles getting on the freeway will be reduced. True, but where are those cars? They are now waiting to get on the freeway. The line of cars backs up into surface streets. When confronted by irate motorists, the project manager replied that this was another project. Yet, both the freeway and the roads and streets are part of one system.

This example still occurs around the world and shows how an organization can establish separate projects to achieve accountability and then create a foundation for more severe problems. It is important in projects to carefully define the scope as well as the objective of a project.

SETTING UP MODERN METHODS FOR PROJECT MANAGEMENT: THE CONSTRUCTION OF THE FIRST NUCLEAR SUBMARINE

Admiral Hyman Rickover was the overall project manager and driver in the construction of the first nuclear submarine. He and his supporters had to face severe resistance from the traditional Navy brass and others within the government. He saw early in the project that the project was not only politically complex, but also technically complex. No one had ever built a nuclear vessel before of this size, much less a nuclear submarine that could remain submerged for long periods of time.

There were many contractors and subcontractors in the project. These efforts had to be coordinated and managed. A number of the contractors kept pressing for resources and money. The problem arose as to how to allocate scarce resources to critical parts of the project. This was a manual nightmare without any methods. Computers were not widely available at this point to support this. Even if they were, there were few methods. Out of this effort came what we know as the Program Evaluation and Review Technique (PERT) and critical path method. Allocation of resources was based on whether the requestor was on the critical path. The relationship between parts of the project was shown on the PERT chart, which we will explore later.

This project is remarkable not only because it produced nuclear submarines, but also because it gave us a set of tools and methods to manage complex projects. The methods were then successively used on other projects and today are now part of most project management software packages that can be purchased for under $500.

A PROJECT WITHOUT A CLEAR OBJECTIVE AND SCOPE: A DOWNTOWN PEOPLE MOVER

We have mentioned successful projects up to this point. This project was started in a major western American city. The objective was to move people around the downtown area with an elevated people mover. People movers exist at airports and in some cities. Before the people-mover project, most people walked around downtown, took buses or taxicabs, or drove their cars. As a project, it suffered from several fatal flaws. People could walk free so many would not pay for the short ride after the novelty wore off. More importantly, it was only later in the project that it was discovered that the time to climb up to the people mover, use the people mover, and then descend to the street exceeded the time to walk the distance. Before the project was killed, millions of dollars were spent on studies, right of way, and engineering.

We can draw from this example that a project must make sense in terms of objective, scope, and strategy. It must be carefully thought out. Otherwise, the fatal flaws will eventually emerge and kill the project. This also suggests that we should carefully consider risk in a project in terms of benefit, feasibility, and need.

—————— EXAMPLES OF SUCCESSFUL PROJECTS

A COMPLEX SYSTEM WITH PROJECT RELATIONSHIPS BETWEEN CONTRACTORS: CONSTRUCTION OF AN AIRPLANE

An airplane is a complex project. Because many aircraft of the same type are built at a plant, we actually have a program. Building an airplane requires a large number of subcontractors for a myriad of parts and subassemblies. The work must be totally integrated.

Today these projects rely on electronic tools for project management and coordination. Electronic Data Interchange (EDI) is used for many transactions between supplier and customer firms. Purchase orders, invoices, payment advices, project status reports, and plans are all handled electronically. These tools with effective management allow the airplane to be constructed at lower cost and with greater speed. The Boeing Company, which has built over 1000 747s (each unique), bears witness to the success of the approach.

A PROJECT WITH GREATER BENEFITS: PUTTING A HUMAN ON THE MOON

In the early 1960s, President John F. Kennedy issued a mandate to place a human on the moon. The United States was under pressure from the Soviet Union. The Cold War and the nuclear arms race were hot. There was competition for the support and commitment of the Third World. The heat was on.

We now look back and see that the project objective was met before the end of the decade. There were many complex parts of the project. But a very critical part was getting a computer on board the spacecraft. It was known that humans could not process and handle the information fast enough to do the lunar landing and orbit of the moon by hand. It had to be done by computer.

In 1961–1962, computers were anything but miniaturized. Transistors replaced vacuum tubes, but the circuits were still wired by hand. The possibility of error was very high. Moreover, the weight and size of computers then prohibited them from being on board the spacecraft. If this were not enough, the computers could not stand the pressure and stress of lift-off and reentry.

Several years earlier, two separate engineers had invented the integrated circuit. Although the transistor was immediately adopted by the computer industry, the integrated circuit found much slower acceptance. The application of integrated circuits solved the spacecraft problem. At the time, the computer on board *Apollo 11* was the most compact computer in the world—all due to transistors and integrated circuits. This application along with other military applications such as the *Minuteman II* rocket contributed to modern computers by their push to integrated circuits.

This was a very complex project and required a dedicated project team at NASA and its contractors over many years. It was a technical and management success. The project had a very clear objective and strategy. The scope was well defined. The methods became established and new tools were created. Looking back, we can see that project success was felt more on entire industries than in the actual achievement of the goal.

USING THE WORLD WIDE WEB

There are a variety of search engines for the Web. Use several of these to go after the topics, "project management," "project manager," and "lessons learned." You will find many Web sites with project information. These include government agencies (e.g., NASA), professional societies, extracts and articles on project management, and information from people who specialize in project management. Bookmark the most interesting sites. If you have software that can search the net when you are asleep, then the search is even more efficient. Consider posting questions in chat rooms.

After finding good material, copy and paste the text into a database. This is better than a word processor since you can index the information. In your database add the following data elements: subject, source, date, web site, title, and summary. If possible, access your database and disseminate lessons learned to others.

WHAT THESE EXAMPLES SHOW

One lesson from these projects is that if a project is successful, not only do a number of factors and tasks have to be right, but also the parts of the project have to be integrated and synchronized. Some characteristics of successful projects are as follows:

- A project with a clear objective has a greater chance of success than one that has a fuzzy goal that can be interpreted differently by different people.
- The scope of the project has to fit the objective. Otherwise, the objective is too broad and can never be achieved; or, the objective is too narrow and the project is completed with little impact.
- Any project must relate to the standard organization of the company or agency. Although it is difficult to do projects within a line department, it is even more difficult to establish a totally independent project. The exceptions that have worked have received massive amounts of financial, managerial, and political support.
- Issues were identified early in the project and given proper attention. Left to fester, some projects result in gangrene or death.
- Project teams were relatively small. Large projects were successful only if they were subdivided.

Looking back, the examples from ancient history show interesting and remarkable achievement. But something else is also prevalent. Many of the civilizations did not have formal project management. They intuitively knew or found out the hard way that to succeed with the project one had to work within the existing organization structure.

We also saw the effect when a clear objective and scope were matched and supported by methods and tools. Many cases of failure were because one or more of these aspects failed individually or in combination at one time or across a span of time.

To be successful, a project of even short duration must adapt to changing circumstances. When the project cannot adapt, it is likely to fail or be stopped. The inability of companies to adapt to peacetime conditions is an example.

EXERCISES

Although we are interested in history, it does not take history to uncover interesting projects from which we can learn valuable lessons. Reinforcing what we said earlier, Project Managers should be aware of various activities in the news and elsewhere. Study these as projects and see what can be learned about the project management process. Learning from an example can save a lot of time.

Another exercise that we suggest is that identifying an unsolved problem that is in the news. This should not be some global problem that is too complex. Examples might be construction projects, government policies, and technology. Define the situation and start to create a project to address the situation.

A third activity is to select a technology. This could be computer or communications based or something else. Do not attempt to understand how it works in technical detail. Sit back and think about how the technology could be used and where it would be useful. What would it take to use the technology? Are there some missing parts?

Use a standard database management system and set up a lessons learned database. Include the following data elements: number, title, date created, project learned from, related lessons, description, and impact. Access the web and identify some lessons learned. Move these into your database. Now try to apply these to projects in your company.

SUMMARY

Guidelines for projects have been presented through historical examples. This is an important part of the learning process. What is striking is the wide range of activities that fit under the general umbrella of a project. This is one of the things that makes project management so interesting and eternal.

Another observation is that project management is much more than a group of people working on something according to a plan. It is much more dynamic in terms of organization, external factors, methods, and tools. If we treat projects as static, we will lose.

Projects evolve over time as they are done. This is true in the ancient examples. It is true today. When we work on a project, we gain knowledge and experience. Over time external factors can influence and impact the project. How the project manager and project team respond is very important to the success of the project and the organization. In our examples, the successful projects tended to recognize symptoms of issues in terms of opportunities and problems early and move systematically to address them.

Another concept we can note in the examples is risk. In many of the civilizations, advances occurred when projects were undertaken that held great risk. The projects still went ahead. What happened to the risk? It was and is still present. Knowing the risks, people took steps to minimize the risk and reduce any effect if problems arose.

II GETTING THE PROJECT STARTED RIGHT

4 Project Definition

In this chapter we shall focus on the "what" of the project—how to establish the project. The next chapter will focus on the how. Chapters 6 and 7 address the who (the project manager and the project team).

A STEP-BY-STEP APPROACH FOR DEVELOPING THE PLAN

Technically, a project is composed of milestones and tasks or activities. Tasks and activities are units of work that lead to a milestone. A milestone is a tangible end product resulting from the task. Managerially, a project is a way to attain a short-term objective outside of the current functional or line organization. Using this second definition we would focus on the relationship between the management of the project and the management of functional areas. Merging these two definitions, a project is a set of interrelated tasks leading to the achievement of an overall objective or milestone. This achievement marks the end or completion of the project.

In defining a project we have to pay attention to both the management and technical facets. The method we will present here for defining the project reflects this. We will define the project as a series of twelve steps. Overall, these are shown in Table 4.1. At each step in the process, we have to go back and review the previous steps to see if any adjustments are needed.

Why go through these twelve steps? Why not plunge in and do the work? A shoot-from-the-hip approach will lead to problems, and possibly failure. People will not understand what is transpiring because the project is not defined and explained. There is a lack of management support because management has not formally approved a project or project plan.

Why go through discrete steps? Working your way through a specific series of steps will help build a better project plan through structure. It will also be easier to gain management support when the project is presented in such an organized manner. Let's examine each of the twelve steps.

TABLE 4.1   Steps in Defining a Project

---

1. Define objectives and scope of project.
   This ensures that project stays under control and is understood.
2. Assess business and organization environment.
   This ensures that the project takes advantage of the known technology, other projects, and information, as well as reflecting the realities of the internal organization.
3. Develop strategy for the project.
   Without a strategy, a project can tend to lack focus and be caught up in the detail.
4. Identify the major parts of the project and overall schedule.
   This ensures a top-down approach to yield a complete plan.
5. Define initial budget.
   This is based on experience and analysis in the previous steps.
6. Identify groups and departments who will participate in the project.
   Need to identify the entire cast of the project to build a detailed schedule.
7. Determine methods to be employed in the project.
   Methods are those used for technical work as well as for control and lay the groundwork for detailed estimates of tasks.
8. Define the tools to be used in support of the methods.
   This validates decisions made in step 7 and determines any additional work required to learn the tools.
9. Refine the budget and schedule.
   Knowing which departments are involved and the methods and tools allows more detailed tasks and milestones of the project plan.
10. Identify the project manager.
11. Establish the project team.
12. Develop a detailed project plan.

STEP 1: DEFINE THE OBJECTIVES AND SCOPE OF THE PROJECT

What is the project goal? Sounds simple, doesn't it? If the project is to build a rapid transit system, the objective would be achieved at the end of the construction. Right? Wrong. The project is only successful if people use the system. If you build something that is not used, you fail. But, some might say, that this is someone else's responsibility. Not anymore. Look back in Chapter 1 where we discussed quality and the global marketplace. The situation is changing in that quality and performance specifications of systems and products are being designed in during the project and not retrofitted later. This broadens the project scope and objectives and makes situations more complex. More organizations have to be involved. In our example, marketing for the rapid transit system should be done in conjunction with design and construction. Trade-offs in design need to be assessed with marketing, operations, and ridership analysis in mind.

In our definition effort, the objective is the goal that we hope to achieve at the end of the project. The scope of the project may include (1) time horizon for doing the project; (2) what is to be included in project tasks versus deleted or moved to other projects; (3) overall involvement of the organization in the project.

There are trade-offs. Making the project too small and narrow means that you will have to create a series of smaller projects—more coordination and management problems at a higher level. Making the project too large means that the project may never be completed. It may be unwieldy. The graph in Figure 4.1 shows this trade-off. The horizontal axis is the scope of the project. The vertical axis is complexity. The two curves represent interproject complexity and intraproject complexity. With a small scope and limited objective (on the left), the interproject complexity is great,

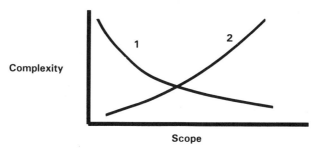

Complexity

Scope

FIGURE 4.1 Scope versus complexity in projects. 1, interproject complexity; 2, intrascope complexity. As scope increases, there are fewer subprojects with less interproject complexity. With reduced scope, there are more subprojects with more interproject complexity.

whereas the intraproject complexity is small. The reverse is true at the other end where the project scope is large. Logically, the best point is the crossover point. However, this is only a theoretical point.

How is the proper scope defined? We first suggest developing a very narrow and a very broad view as alternatives. Now identify what has been deleted when moving from large to small. These would result in additional projects if this work were to be performed. How do you determine if some area is in or out? Here are some guidelines in the form of questions:

- How closely related is the work to the base project?
- Would the same resources be used on the work and the project?
- If they are kept separate, what will be the advantages and disadvantages be?

Advantages to having the work in one project include the following:

- A smaller project and its tasks may be swallowed up by the larger project.
- The project overall is too large to manage.
- The smaller project loses its identity.
- Centralizing risk into one project.

In the past managers tended to combine the work into a smaller number of larger projects. Seeing that this can result in failure, the tendency has been to move toward more, smaller projects and endure the coordination cost among projects to reduce risk. Table 4.2 gives some alternative objec-

TABLE 4.2   Alternative Views of Two Projects

Historic project: Hadrian's Wall in England
The wall was designed with gate posts at every mile regardless of whether they were
   needed. The design was followed rigidly, regardless of terrain.
Alternative 1: Protection
   Purpose: Build wall to keep barbarians in Scotland out.
      Scope: Focus on entire infrastructure including troops to guard the gate posts.
Alternative 2: Monument to the emperor
   Purpose: Build a large structure in honor of the emperor that has some useful purpose.
      Scope: Focus on architecture and structure and not infrastructure.
Modern example: Cement plant in Burma
The Chinese government some years ago constructed a cement plant in Burma (or, today,
   Myanmer). The plant was a failure in that the bags of cement hardened in the humid
   air as they awaited shipment.
Alternative 1: Support the local economy.
   Purpose: Implement a modern plant that could open industry in the country.
      Scope: Focus on entire infrastructure, including the plant.
Alternative 2: Be a political showcase.
   Purpose: Be a successful demonstration project.
      Scope: Focus on construction and process and not on end products.

tives and scope for a historical and current example. Table 4.3 illustrates ten different interpretations of purpose and scope for one project.

What happens if you make an error? Typically, you will find out about it later, after the project has started. Things will not be going right. There may be management and coordination problems. What do you do? We will address this in detail later, but the answer is to implement systematic changes.

Let's consider another historical example. We are all familiar with the travels of Marco Polo to China. The goal of his travel was to trade and to open up China for his native Venice. The objective was defined. The scope

TABLE 4.3 Alternative Purpose and Scope for a Project

The example is the Concorde supersonic jet aircraft. The final cost for a small number of planes was over 400% of the original estimate. The original estimate was made in 1959. The plan was certified in 1975. Production ceased in 1980. Moreover, the plane was restricted to specific cities due to the sonic boom.

Alternative 1: Technical, political
    Purpose: Demonstrate to the world that Europeans can build advanced technology.
      Scope: Focus on design and demonstration.
Alternative 2: Technical
    Purpose: Demonstrate that commercial supersonic flight was feasible.
      Scope: Focus on design and development.
Alternative 3: Cooperation
    Purpose: Demonstrate that Great Britain and France could cooperate on a major project.
      Scope: Focus on schedule and management.
Alternative 4: Technology management
    Purpose: Demonstrate ability to manage complex, technical project.
      Scope: Focus on technical and production aspects of project.
Alternative 5: Economics
    Purpose: Make a profit
      Scope: Focus on schedules and production as well as efficient design.
Alternative 6: World leadership in commercial aircraft
    Purpose: Achieve a major technological breakthrough.
      Scope: Focus on engineering.
Alternative 7: Support local industry
    Purpose: Subsidize through the project the aircraft industry in Great Britain and France.
      Scope: Focus on internal engineering and production.
Alternative 8: Start new industry
    Purpose: Use Concorde as start-up for a new industry.
      Scope: Focus on the industry infrastructure.
Alternative 9: Consumer orientation
    Purpose: Provide fast, comfortable travel.
      Scope: Focus on design and affordability.
Alternative 10: Political
    Purpose: Provide support for the government in both countries.
      Scope: Focus on jobs versus the project itself.

was broad. The travels were successful. But have you ever heard of Ibn Battuta who lived during the fourteenth century? He was raised in Tangier in northwest Africa. Over almost 50 years, he traveled more than 75,000 miles across all of the Moslem world. He traveled across Africa, the Middle East, western Asia, and even India and Sri Lanka (Ceylon). His purpose was to explore the Islamic communities and their faith. His writings and even his name are largely unknown except to scholars. What is the difference here? Battuta had a clear objective, but his scope was narrower. He was successful, but because of the narrowness of his project, his work and travels are largely unknown and have had limited impact.

STEP 2: ASSESS THE ORGANIZATIONAL AND BUSINESS ENVIRONMENT FOR THE PROJECT

This is often ignored if overlooked. Having defined the objective and scope, we define the boundaries of the project and understand the external environment surrounding the project. Why do this? Why not just start by defining more of the internal detailed characteristics of the project? Because the project will be impacted by external factors. Also, we need to validate the objective of the project. The achievement of the objective of the project will result in some material change within, and perhaps, outside of the organization. Understanding the external environment helps to validate the objective.

Fine, now what do we do and how do we do it? After all, understanding the environment is vague. We begin with our organization. We ask the following questions:

1. What parts of the organization will be affected by the project and by the end product of the project? How will they be impacted? The answer leads us to consider whether the scope of the project should be modified to accommodate any other groups.

2. What is the relationship between our project and other projects? Are there dependencies in work? Are there shared resources? Do they have methods and tools in common? Answers to these questions will assist in determining whether our project should be modified to relate to other projects.

3. How do external factors such as customers, technology, suppliers, regulations, politics, and other forces impact the project and its end product? The answer to this question impacts both the scope and objective because the end product of the project relates to the overall project purpose.

Understanding the external environment means that we are acknowledging the political realities of the organization and its environment. What happens if we skip this step? We are more likely to wind up with a project that cannot be completed because of the unavoidable impact of an environmental factor on the project. Another trap is that the end product of the project will not be used properly, if at all.

Table 4.4 gives a list of environmental factors. Consider the building of the Roman Coliseum (which we will revisit at the end of this chapter). We could interpret the project as the building itself. The scope would then logically be the support and management of the construction. This is wrong and it does not reflect what really happened. The Romans thought it through. They saw that if they were going to stage wild animal hunts and other events, they would need to have a logistics network from Africa to support it. History shows that this logistics network was related to the shipment of grain and other foods from Africa to Rome. Northern Africa was politically and militarily stable and produced substantial food-stuffs for Rome. Northern Africa was important because the Roman diet depended on grains and other similar products as opposed to meat.

Bringing us up to the present, here are some suggestions that we usually consider in evaluating the environment:

1. Because we are going to do the project, what can we do with little incremental cost to increase the benefit and impact from the end product? Through a modest expansion of scope, we may have a substantially increased benefit.

2. Do not accept the stated objective as a given. What are related changes that can be done at the same time? Analysis of the environment will identify other areas that can be addressed as part of the same project. Alternatively, we can look for follow-up projects that can be carried out immediately after our project.

3. If the project is too big, especially in terms of time, then you should consider the possibility of having interim milestones and subprojects. This raises another basic issue: Are there interim milestones and end products in the project that have value in and of themselves? Achieving early successes in the project will increase credibility and possibly muster more support for the project. It will also allow for refinement and change in the project. This is what happens when an architect builds a scale model of a structure. People see the model and get excited and give more support to the project. This is an idea that goes back to ancient Egypt and China.

At the end of this step, we have defined the environment and identified the elements that will impact and be impacted by our project. We will have refined our objective and scope. This will help us in later steps when we identify organizations that need to be involved in the project.

STEP 3: DEFINE THE STRATEGY TO BE USED TO ACHIEVE THE OBJECTIVE IN LIGHT OF THE REALITIES OF STEP 2

We have a project objective that is a general goal and relates to the end product. We also have the scope of the project. But how do we accomplish

TABLE 4.4   Potential Environmental Factors

A. Technology
   1. Available technology
   2. Gaps in current technology
   3. Technology likely to be available in near term
   4. Long-term trends in the technology and providers
   5. Technology support requirements
   6. Interface capabilities between technologies
B. Competition
   1. Technology used by competition
   2. Organization structure of competition
   3. Business processes and how they are managed or structured
   4. Products and services offered
   5. Delivery system used by competition
C. Regulation
   1. Taxes and tax structure
   2. Restrictions from certain businesses
   3. Government review and approval process
   4. Reporting requirements
   5. Health and safety regulations
   6. Labor laws
D. Social
   1. Available pool of workers
   2. Attitude of society toward company
   3. Potential impact of project on society
   4. Attitude of society toward project
   5. Security
E. Infrastructure
   1. Transportation system
   2. Health and safety
   3. Utilities
   4. Communications
F. Industry
   1. Current practice in industry
   2. Trends in practice in industry
   3. Competition from other industries
G. Customers/suppliers
   1. Potential impact on project
   2. Benefits of project to customers or suppliers
   3. Involvement in project
H. Parent company/subsidiary
   1. Priorities and focus of company
   2. Importance of project to the company
   3. Dependencies with the project
   4. Involvement with the project
I. Labor unions/organizations
   1. Current labor agreement
   2. Work rules
   3. Flexibility to change work rules and business processes
   4. Attitude of labor toward project
J. Political
   1. Political attitudes toward company and project
   2. Political impact and fallout from project
   3. Cross-impact on other projects

this? We need a general strategy for the project. What is a strategy? It is a general definition of the resources, methods, and tools to be used in the project. Here are some examples of strategies:

1. Work on the project when the people are not busy with their normal work. This is how the pharaohs built the pyramids—employing people in the off season. It is also how the canals were constructed and maintained in the Tigris-Euphrates Valley of the Middle East.

2. Employ automation and automated testing to the fullest extent. This is how many companies have established new production lines. Motorola, for example, followed this approach with building its pagers.

3. Use no internal employees except for management and coordination. This is how governments contract out for constructing highways and other public facilities.

We have to be careful with strategies. Like objectives they need to be flexible so that when an objective changes, the strategy can change. Often, the strategy will change while the objective remains the same. Haven't we all been told, "Well, you'll have to do the same thing in less time (with less resources)"? Same objective, but altered strategy. What is a rigid strategy? Examples are ones that lock us in with a specific method or tool or resource. You bought a hammer for the home project. But the hammer is not the right tool. You don't have any more money for more tools. Do it with or without the hammer, but that is all you get. How do we construct a strategy? Well, you have the objective, scope, and environment. You know the general time frame. What suggestions come to mind when considering resources? What organizational issues come to mind that should be included in the project? A strategy should provide the basis for some of the following:

- How you organize the project
- Selection of the project manager
- Which organizations should be involved
- Types of people required for the project team
- Methods to be used in the project
- Tools to support the methods

How should you develop a strategy? First, as with the objectives, you should consider the extremes. At one end is nothing—no strategy. At the other end is a detailed strategy that is far too confining. The right answer is somewhere in the middle. Sit back and ask yourself where the risk is in the project. Where can things go wrong? Target your strategy to address the risk while still maintaining flexibility. Table 4.5 gives some examples of strategies for specific objectives.

STEP 4: DETERMINE THE MAJOR PARTS OF THE PROJECT AND AN OVERALL SCHEDULE FOR THE PROJECT

You now have enough to start determining the major parts of the project. You can also think about the general schedule so that the budget can be

TABLE 4.5   Examples of Strategies versus Objectives

This example uses the alternatives of Table 4.4 for the Concorde supersonic jet aircraft.
Objective: Demonstrate to the world that Europeans can build advanced technology
  products.
 Strategy: Focus on the project as a technology project rather than an economic project.
Objective: Demonstrate that supersonic flight is feasible.
 Strategy: Focus on product feasibility
Objective: Demonstrate that Great Britain and France could cooperate on a major project.
 Strategy: Demonstrate publicly cooperation; manage in a cooperative way.
Objective: Demonstrate ability to manage a complex, technical project.
 Strategy: Focus on project management
Objective: Make a profit.
 Strategy: Organize work to achieve low cost, on-time results; deemphasize politics and
  excessive new technology.
Objective: Achieve a major technological breakthrough.
 Strategy: Focus on technology over a 20 or more year time horizon.
Objective: Subsidize through the project the aircraft industry in Great Britain and France.
 Strategy: Spread the work on the project through the industry.
Objective: Use Concorde as a start-up for a new industry.
 Strategy: Expand scope of project to include the industry and industry structure.
Objective: Provide fast, comfortable travel.
 Strategy: Narrow strategy on the design in terms of passengers.
Objective: Provide support for the government in both countries.
 Strategy: Expand project to include political structure and cooperation.

developed. You do not want to establish the detailed plan at this point. Why? One reason is that if you are not the project manager and you define the plan, the eventual project manager might not assume ownership of the plan. You want the project manager to commit to both the schedule and plan. A second reason is that even if you are the project manager, you should develop the project plan top down. If you develop the plan in detail and later discover more information about the project, then you will have to make several major revisions.

Major parts can be phases or may involve a certain percentage of the estimated overall effort. Here are some thoughts on how to define the parts:

- Divide the project by natural phases (e.g., analysis, design, building, testing, etc.).
- Divide the project by time periods.
- Divide the project by major milestones.
- Divide the project by the type of work to be performed (this might be on organizational lines).

The first and third ways of dividing are preferred because they are dependent on the nature of the project itself. Dividing by time period is artificial. Dividing it by type of work will often require too much detailed analysis. Remember that you are doing this at a general level.

Do not let one phase or part be much bigger than all of the rest. Otherwise, you will likely run into problems when the schedule is developed with one large section. Have at least one milestone for each part. You should also evaluate several alternatives. One simple approach is to attempt to quickly divide the project into five parts and then ten parts. This will render a consistent decomposition of the project.

STEP 5: DEVELOP AN INITIAL BUDGET

You will have to give management some overall range or idea of what it will take to do the project. Obviously, with the first four steps this cannot be very precise. What is reasonable is to define the areas or categories of the budget (personnel, equipment, etc.). Given the relative size of the pieces of the project and the schedule, you can then estimate the resources. You will obviously want to be conservative in your estimates. Include all expenses. Examples of commonly omitted expenses are acquiring tools, getting training, documentation, and other supporting activities. Another error is that steps are left out and this leads to the budget being underestimated. You need to allow for other contingencies as well, such as

- Change in project scope
- Management placing new requirements on the project
- Management withdrawing some of the resources from the project or project is delayed
- Management changing the priorities of the project
- Competitor doing something that impacts the project focus

When you put this budget together and present it, it should be done with the assumptions that you made when you prepared the strategy and budget. These assumptions relate to the availability of people and other resources and the availability of work from other interdependent projects. Other assumptions might relate to the scope of the project.

STEP 6: IDENTIFY THE DEPARTMENTS AND GROUPS THAT WILL BE INVOLVED IN THE PROJECT

From the environment in step 2, the strategy in step 3, and the overall parts and schedule of step 4, you can start identifying the organizations that will have to be involved in the project. You should err on the side of including organizations and groups as opposed to excluding some.

Include management as well as the organizations. What will these organizations do? They may participate in the project. They may have projects or work that will serve as input to the project. They may be involved in evaluation of project milestones, or they may serve as bases for technical support. For each organization, write down what you view as their role.

This information is necessary for several reasons. First, it validates the scope of the project. Second, all parts of the project can be mapped against the organizations you have identified. Third, it will help you later when you gather the project team together and set up the detailed schedule.

STEP 7: DELINEATE THE METHODS TO BE USED IN THE PROJECT

Methods apply to project control as well as techniques for organizing and doing the work. Why do you need to identify the methods now? The methods will serve as a framework for how the work will be done on the project as well as how it will be managed.

It is useful to divide the methods into categories. Some categories we have employed are as follows:

1. Methods to control the project and work. This covers how problems will be handled, how project updates will be done, how reviews of milestones will be conducted, how budget and schedule will be reported, and other control aspects of the project.

2. Presenting and dealing with management. This encompasses how the project manager and project team will interface with upper management, management of other projects, and functional managers.

3. Doing the work. These are methods that will be used to perform work in the project.

This sounds simpler than it is. In some projects where people come from different organizations to work on the project, they may bring with them the methods of their home organization. The project manager has to clearly identify the methods that will be employed at the start of the project.

STEP 8: ASCERTAIN APPROPRIATE TOOLS TO BE EMPLOYED TO SUPPORT THE METHODS

A tool can be automated or manual or both. Consider a shovel to dig a hole. The shovel is the tool. How you will dig the hole is the method. Keep in mind that we have defined the methods first in step 7. Many people tend to start with the detailed tools because they are more tangible. The project then runs into the danger of being tool driven. When the tool doesn't work or doesn't fit, there are difficulties. On the other hand, if the tool is supporting the method and it fails, then the method is still intact.

STEP 9: REFINE THE BUDGET, SCHEDULE, AND SO FORTH

We highlight this as a distinct step because it is logical at this point to go back and review the work of previous steps. Remember that all of the step results have to be consistent with each other. For example, the scope should match the overall schedule and budget. We suggest here that you attempt to reduce the results of the steps into lists or tables. If you write text, then it will be more difficult to match these up between steps.

STEP 10: SELECT THE PROJECT MANAGER

This is the subject of Chapter 6. When we discuss the characteristics and duties of the project manager as well as their activities, we will also discuss the attributes of how we select a project manager. It is not generally true that you want to pick a seasoned veteran. If the project is novel, then you might pick someone with initiative and drive as opposed to experience. A basic observation we can make based on our experience is that you are more successful if you select a project manager based on personal attributes rather than detailed experience. You might pick project team members based on expertise or experience, but you pick a project manager based on his or her ability and willingness to manage and deal with issues and conflict. Note that project leaders can be changed to accommodate different phases of a project. Someone who is good at leading a design effort may not be suitable for development and implementation.

STEP 11: BUILD THE PROJECT TEAM

We will also deal with the project team in a later chapter. If you have identified the organizations that will participate and these other steps, it should be simple to build the project team, right? Again, wrong. You have to sell the project to the manager of the functional department. You should not ask for a specific person unless they inquire if you have someone in mind. You should describe the project and what you need from them. Then follow the suggestions of Chapter 7 in getting good team members and in getting them on board.

STEP 12: DEVELOP THE DETAILED PROJECT PLAN AND REFINE THE BUDGET

What is in the detailed project plan? It includes all tasks, dependencies between tasks, resources, milestones, and schedules. It includes costs, an identification of potential risks, and contingencies. It also includes the general part of the plan related to objective, scope, strategy, and environment. The detailed project plan is usually composed of text, charts, graphs, and tables. We examine specific methods in Chapter 5. The text should not change much over the project because it is the basis for the work.

Why build the detailed project plan as the last step? Because you need the information and results from the previous steps. Also, with a project manager and project team, you can get them to participate in the development of the plan and schedule. They can validate the plan for completeness. With this consensus there are likely to be fewer problems later. Also, you will be more likely to obtain management approval for the project.

Where do we start? Having defined the general parts of the project, list these and start to build a set of major milestones for the project. After you run out of steam in generating these milestones, move to defining the tasks. The set of tasks and milestones is what is referred to in project management as the Work Breakdown Structure (WBS). As we will see in Chapter 5, it is useful if you can use an existing work breakdown structure as a starting point. People will know what the tasks mean and management already is familiar with them. But a canned structure will not fit all projects. Do not be afraid of creating or modifying a structure to fit the project.

How should you build the tasks? You could get the project team together and ask them to do it. This will be very time consuming if the project team is very large. Moreover, people might work on the same tasks. Therefore, we suggest that you, as the project manager, develop an initial set of tasks and milestones. We call this a strawman schedule. Distribute the strawman to the project team and explain the tasks and milestones to them. Ask them to go and make additions and changes. When they have worked on this for some time, go to each one and get their comments and suggestions. With all of the changes, create a new list of tasks and present the revised list to the project team. Note that the project manager has shown leadership through this process. Also, note that the project team has begun to get involved and participate. This is less threatening because no tasks have been assigned. No schedules have been set.

Now that the tasks are built, start identifying dependencies between the tasks. You are not setting up schedules yet, but you are attempting to determine what tasks depend on others. Start by getting clear dependencies. A clear dependency is one in which A must definitely precede B. If there is doubt, leave them as parallel tasks. Your goal here is to have a minimum of dependencies, but you also want to identify those that are required. Why a minimum? Because any dependency may create a problem in tasks not being worked on and because a large number of dependencies can make the schedule more unmanageable.

Once you reach agreement on the dependencies, get the project team together and start identifying a specific organization (not a person) that will be accountable for each major task. Consider what we have just said. We said organization and not person because people on the project team can change. You are after organization accountability. The project plan will have to be presented and sold to the managers of the project team.

We have focused on a single organization. For some tasks there may be several organizations or groups involved. You should name the major organization. The person from that organization can later identify any other organizations that have to be involved in terms of support.

The third thing we mentioned is a major task. If you have a list of 200 tasks, then addressing each one in a meeting is too time consuming. Divide the tasks into groups. Perhaps, you might use the major project parts that we developed in an earlier step.

With the tasks and milestones defined, and the dependencies and accountability established, you can now assign the tasks to the relevant project team members. Ask them to get back to you with a schedule for the tasks for which they are responsible. After you get the information, then you can build the schedule and plan. You can now refine it with the project team and get consensus.

We also have to refine the budget and schedule based on the detailed plan. We should also revisit the methods and tools that will be used in the project to make sure that we did not omit anything. We also indicated that we must delineate the project risks. An event in a project is where something can go wrong. The event carries with it a likelihood of occurrence. If the event occurs, then there will be cost of recovery and contingency. This is is called exposure. Mathematically, risk is the product of likelihood and exposure. If you don't want to deal with the mathematics, then make a list of what can go wrong. Start to rank these by the likelihood that they can occur. Also, rank them by the damage that would occur. For the highest ranking events, start thinking about contingency plans and what can be done to recover.

We will address risk in more detail in Chapter 9. Here we will mention several examples. One is loss of a key project team member through reassignment. A second risk is the unavailability of information or an end product from another project. There are many others, but this should give you the general idea.

THE PROJECT TEMPLATE AND THE WEB

Can we abbreviate or at least reduce the effort to create the plan? Yes. The traditional approach was to have people select tasks from a predefined list of, perhaps, thousands of tasks. This list was the Work Breakdown Structure (WBS). Today, with modern project management software on a network, we can do better.

We can standardize on the following:

- High level summary tasks (the most detail would be general tasks of one month or more);

- Dependencies between the high level tasks;
- List of resources (personnel types, equipment, facilities, etc.);
- Assignment of resources to tasks.

Taken together we have a template. We can store the template plan on the internal network or the web. Project leaders can then access the network and download the template as a starting point—saving time. A second benefit is consistent project structure to permit analysis and lessons learned across multiple projects. The project manager retains flexibility since detailed tasks and durations will be added by the project leader.

We maintain several lists of resources—generic and specific. Generic is used for the template and specific is used by the project leader. For example, test equipment is generic and test equipment no. 1 is specific. When setting up a schedule the project leader replaces the generic with the specific resources.

The template can be established on the Web. Tasks can be defined and explained in web pages as can lessons learned. We have found that linking experience to tasks in the template helps encourage the use of the template.

PRESENTING THE PROJECT PLAN

In Chapter 12 we will discuss presentation methods. Here we place our attention on the overall structure of the presentation. There is a tendency to plunge in and present the results of the last step—the detailed project plan. After all, this represents the culmination of the work in the previous steps. We suggest that you not do this. Instead, we suggest that you divide the presentation into two parts. The first part is for the general approach. This consists of steps 1, 2, 3, 4, and 6. This provides an overview to management of what the project is and what you hope to accomplish.

The second part of the presentation deals with the detail. Now present step 12 along with the budget and schedule. This is the detail. It is far easier to sell the detailed steps if the audience knows what the general picture is.

When you are presenting the general part, you will often get few questions. The reason may be that they are waiting for the detail. Then when you present the detail and the schedule is too long or budget too high, you may get questions. The answers to the questions will often lie in the general part of the presentation and plan. Management is seeing, for perhaps the first time, the full impact and requirements for the project.

What will happen as a result of the presentation? You may be asked to revise the schedule. Changes should be based on changes to the general plan (strategy, scope, etc.). You should change this first and then change the detail.

EXAMPLE: COLISEUM OF ROME

This is an elaboration on history and illustrates political, social, techno-logical, and other elements that made it a complex project. Note that we have taken the wider scope to include logistics and politics. The environment step is very brief and deserves some additional comments. Roman citizens were a favored lot. They had many more holidays than we have today. Food was provided to many. When rulers and generals achieved victories, the public expected that public buildings would be erected in honor of the victories. To gain public support, many Caesars provided games and contests as entertainment—the classic bread and circuses routine. The emperor of the time, Vespasian, was no exception. One reason that the Coliseum was so successful was because it met multiple social needs.

EXAMPLE: KGB PROJECT TO PLACE MISSILES IN CUBA IN 1962

One of the greatest crises of the Cold War occurred in 1962 when the Soviet Union placed over one hundred nuclear armed missiles in Cuba. After the face-off between the United States and the Soviet Union, the missiles were supposedly dismantled. How did the missiles and over 40,000 scientists and soldiers get into Cuba in the first place and then for months deploy the missiles undetected? This is the story of one of more audacious and successful projects in the 1960s.

The KGB, as was learned from recently unearthed archives, developed a detailed project plan. The objectives and scope were very clear. Diversionary project plans were established. The code name of the project referred to an Alaskan location. Soldiers were not told of their true destination until they were at sea. They had taken winter clothing, thinking that they were heading for the Arctic. Instead, they were headed for Cuba, ninety miles from the United States. The project plan included the management of over 100 ships disguised as freighters. The soldiers changed into casual clothes and were greeted in Cuba as agricultural advisors. They then proceeded to systematically move and establish the missile bases throughout the island with the various support facilities in terms of radar, troop barracks, and so forth. The project of setting up the missiles was successful and went undetected for months. This example shows the result of focusing on specific objectives with a defined scope and then devoting the necessary resources to achieve the objectives.

EXERCISES

We have two exercises. The first one is inductive. That is, take a small project that you are working on or will work on at home or at the office.

Try to work through the steps. You will have some trouble if this is the first time. There is a tendency to just make a list of tasks and start at step 12. Resist this and sit back and think about the project. For example, if you are going to add some furnishings to a room, think about your house, apartment, or condominium in general. Where should you make improvements? What will it do for your lifestyle? Many people buy things because they are on sale and because they fill a gap. But you may discover more important needs as you think about the general picture.

The second exercise is deductive. Take a project with which you are familier. You have a task plan in front of you. Work backwards through the steps. Some steps such as naming the project team and organizations will be easy. Other steps will be more challenging. For example, can you detect the overall strategy from the plan? What is the scope of the project? If either or both of these are not clearly evident, then this may be a sign of potential problems with the project. What can you learn from this? You see the importance of having the scope, environment, and strategy defined. If you see that the project plan changed, then you should be able to detect a shift in strategy or scope of the project.

Take a project plan and see if you can generate a template. Extract and summarize high level tasks and make the resources generic. Then eliminate the detailed tasks and set up dependencies among summary tasks.

GENERAL EXAMPLE

The example we discuss here will be covered at the end of each chapter. A large manufacturing firm wishes to reengineer the way it produces its basic product. The product is complex and composed of a number of subsystems that have to be integrated and tested prior to delivery. There are a number of problems with the current process. First, the project management process to plan and track products is distributed in such a way that there are over ten active project plans—no two of which agree. Second, the actual details of some of the work in production are not documented in depth so that it is difficult to determine status. Third, as with many organizations, there are conflicts between organizations and blame when production schedules slip. Different groups employ different tools and methods in doing project management, thereby compounding the difficulty in getting an integrated schedule that crosses all departments. Some of the software tools are complex to use and require almost programming skills. This in turn requires a number of intermediaries between the line managers and upper management in doing project management. The intermediaries use the software and attempt to track the production. However, schedules are often out of date because the intermediary is not involved in the production process. Upper-level management then in turn distrusts the schedules.

Obviously, there are many projects here. They range from the production process itself to the project and product management process. Our overall goal is to define and implement a project management process that addresses the issues defined above and also contributes to the basic improvement and reengineering of the production processes.

Key issues are where to begin and how to establish the purpose and scope of the project. If we define the project to be reengineering of the entire process, then the scope is too broad and the project will fail before it shows results. If we define it too narrowly in one division or department that is not critical to the overall manufacturing process, then it may not be considered significant and representative. We do know that we need to address the project management process first somewhere. This will lead to a better understanding of the production processes and contribute to reengineering. But, two key issues remain. What should be the overall strategy? Second, where should we begin?

Let's consider the strategy. We cannot attack all departments at one time. We need to carefully choose our first department and then with success select related departments for more work. We need to be flexible in order to take advantage of the dynamics of the project. Some of the dynamics are that if the new project management approach works, different departments will evince varying degrees of enthusiasm. We do not want to delay an enthusiastic department based on a strategy developed six months earlier. Thus, our strategy is one of flexibility and, if resources permit, to stage different departments to allow for parallel effort.

In the real life example, the critical area of integration and final assembly was chosen as the first area. This was an area where there had been substantial slippages in the past and where improvements seemed to be possible. However, it was also an area of risk because much of the production work had already been done prior to final assembly. Using our project strategy, after this area was underway, less risky and earlier phases of manufacturing would be addressed successively.

This example indicates that it is useful to have an overall strategy that defines a series of projects with their own objectives and scope within an overall objective. It also shows that to be successful, you often must take risks up front in the first project. Another lesson is that there is real benefit in maintaining flexibility.

Before leaving this example, let us highlight some of the twelve steps with comments.

Step 1. The overall objective is to improve the project management process; the scope of the total project is the manufacturing organization.

Step 2. The organization and business environment was already summarized. The project must be sensitive to these political realities.

Step 3. The strategy has been defined.

Step 4. The major parts of the schedule for the initial project involving integration and final assembly include understanding the current process, getting involvement and participation from line managers in project management, pilot implementation of the new project management approach, demonstration and refinement of the project management approach, and implementation of the new project management process on a production basis.

Step 5. The budget will be established in detail on a phased basis in line with the phases in step 4.

Step 6. The departments involved depend on the focus of the individual project.

Step 7. The methods to be used focus on line manager and supervisor maintenance of their schedules and the use of the scheduling methods and tools on a daily basis.

Step 8. The tools involve electronic mail, electronic forms, microcomputer project management software, and database management software for schedule integration using a client-server approach.

Step 9. Refinement was accomplished with management participation and by holding group sessions that identified and focused on issues.

Step 10. The project manager was chosen to be someone who had extensive hands-on manufacturing experience as well as project management experience.

Step 11. The team included several full-time people, a former employee, and a consultant. Other people in the project entered and exited by phase.

Step 12. The detailed project plan will be discussed later.

SUMMARY

We have outlined twelve steps that should be addressed in building a project plan. Most of these steps have to be performed in sequence. For example, defining a strategy should follow the objective and environment. But there are some parallel activities. The identification of organizations can be done in parallel with the first estimate of the budget, for example. The importance of a template cannot be underestimated.

5 Setting Up the Project Plan

In this chapter we concentrate on the details for setting up the tasks, milestones, and schedule. Beginning with a static view of the project in terms of defining tasks, resources, dependencies, and graphs and charts, we will move to scheduling—the dynamic part of the process. Scheduling is the process whereby we attempt to fix when tasks will be done and when and how many resources are required for these tasks.

TASK DEFINITION

What defines a specific task? The name should be short and unambiguous. Because many tasks sound similar, tasks should be numbered. Also, because some tasks are parts of larger tasks, devise a numbering system that reflects this. For example, consider

1000   Plant flowers.
1100   Prepare the soil.
1200   Select and purchase the flowers.
1300   Plant the flowers.

This trivial examples shows that there are three major tasks under task 1000 Plant flowers. This is indicated by the task numbers 1100, 1200, and 1300. Why have all of the zeros? What if you want to add more detail? It would be hard to do if the tasks were numbered 11, 12, and 13. But with the zeros, we can add the next level of detail, as follows:

1000   Plant flowers.
    1100   Prepare the soil.
        1110   Get tools.
        1120   Dig up soil and remove weeds.
        1130   Prepare the soil.
    1200   Select and purchase the flowers.
        1210   Select the flowers.
        1220   Identify fertilizer.
        1230   Purchase the flowers and fertilizer.
        1240   Take them home.
    1300   Plant the flowers.
        1310   Prepare the ground with fertilizer.
        1320   Plant the flowers.
        1330   Water the flowers.

Note that the tasks start with a verb. Also, we could still add one more level of detail without changing the numbering system. You may wish to use a different system, but adopt some consistent system that allows flexibility. What happens if you have to renumber the tasks? People will think that the plan is being changed. It may take some time to reexplain all of the tasks. You think we are kidding? In one project the manager renumbered a 250-task project plan. It took over an hour to explain it and convince everyone that only the numbering had changed.

Note that we did not indent the tasks as in outlining. A work breakdown structure is a list of tasks in an organized structure. It is an outline. However, if we indent, then we lose space. The space in a manual form as in a computer project management system is limited to a certain number of characters for the name. That's why we discourage indenting.

How should you develop the list of tasks? Top down. Start with the major parts or phases of the work. Include overhead tasks as well. What do we mean? Table 5.1 gives some examples of overhead tasks related to reviews, project teams, training, and project management. Err on the side of includ-

TABLE 5.1 Examples of Tasks

A. Methods
1. Evaluate methods.
2. Select methods to be used.
3. Define measurement of methods.
4. Train material to support method.
5. Train cases and data.
6. Plan training.
7. Develop guidelines for training.
8. Determine training audience.
9. Conduct pilot training.
10. Evaluate pilot training results.
11. Conduct training.
B. Tools
1. Review methods.
2. Identify potential tools.
3. Evaluate tools.
4. Select tools.
5. Determine tool use in support of methods.
6. Define how tool will be used in general.
7. Define training needs for tool.
8. Identify tool expert.
9. Train the tool expert.
10. Plan pilot project using the tool.
11. Carry out pilot project.
12. Review results of pilot.
13. Develop guidelines for tool use.
14. Plan for training in tool use.
15. Define audience for tools.
16. Conduct training.
17. Develop methods for measuring use of tool.
18. Undertake tool review.
C. Project reviews
1. Define purpose of the review.
2. Determine scope of review.
3. Identify documents and systems to be reviewed.
4. Determine method of review.
5. Identify reviewers.
6. Define reviewing schedule.
7. Conduct preliminary planning for review.
8. Distribute materials and guidelines for review.
9. Perform review.
10. Conduct review meetings.
11. Document review results.
12. Present results of review to management.
D. General overhead tasks
1. Arrange for facilities.
2. Coordinate transportation.
3. Arrange for supplies.
4. Input data into project management software.
5. Identify potential contingencies for project.
6. Update project plan.
7. Develop alternative projected project plans.
8. Analyze projected vs. actual vs. planned schedules.
9. Review progress with each team member.

ing tasks even if you think they may not apply. It is unpleasant to add anticipated tasks.

When you are starting out, look for canned, existing lists of tasks or work breakdown structures in your organization. These are already known to people without training or explanations. This will give you standard names and abbreviations as a starting point.

You should use standard abbreviations wherever possible because the length of the name is limited.

MILESTONES

Technically, think of a milestone in the plan as a task that has no length or duration. Milestones should be numbered like tasks. How do you tell them apart from tasks? Another suggestion or two. You can number these with a "9" as the last digit in the number. And start it with the letter "M."

In our example, we have several milestones: ground is ready for the flowers, flowers are ready to plant, flowers are planted. If we insert these into our plan, we have:

1000   Plant flowers.
    1100   Prepare the soil.
        1110   Get tools.
        1120   Dig up soil and remove weeds.
        1130   Prepare the soil.
        1199   M: Ground ready for flowers.
    1200   Select and purchase the flowers.
        1210   Select the flowers.
        1220   Identify fertilizer.
        1230   Purchase the flowers and fertilizer.
        1240   Take them home.
        1299   M: Flowers ready for planting.
    1300   Plant the flowers.
        1310   Prepare the ground with fertilizer.
        1320   Plant the flowers.
        1330   Water the flowers.

See how the milestones stand out in the list? Also, note that the milestone starts with a noun. In developing the words for milestones, follow the suggestions that we gave for tasks. Try to keep adjectives and any political words out of the plan and try to use clear and simple words.

Also, note from the example that there is an overall task followed by subtasks. There are several milestones that correspond to major phases or parts. There is not one milestone for each task. This is also common practice.

IDENTIFYING RESOURCES

Before going into more detail on tasks, let's look at resources. We said before that we will define a list of resources and then apply these to the tasks we have defined. What is a resource? Here are some categories.

- Organizations that will participate in the project
- Specific individuals who will participate
- Functional roles that will be needed in the project (examples are management, line management, etc.)
- Equipment
- Furniture
- Facilities
- Test laboratories
- External contractors
- Other projects
- Utilities

These resources are very different from each other. Some do the work; others are needed to do the work. Most of the items on the list fit these two general types. "Other projects" is included because another project may provide input to your project.

Let's consider how to identify and name these resources as we did with tasks and milestones. We should have a short name for each resource. The reason is that the short name will be more useful for manual and computer reports. We also have a full name. We suggest standard abbreviations and avoidance of punctuation.

What can we say about a resource? First, we can determine its availability for the project. We could also assign costs to a resource. Costs are fixed, unit, or variable. Fixed costs are resource costs that do not depend on extent of use. Unit costs apply to resources where the more something is used, the higher the cost, but at the same unit of cost. Variable costs allow for discounts based on more use. We can also determine whether the resource can be leveled in terms of being spread out if it is overallocated.

Supporting a resource is a calendar. Usually, there is one work calendar for the entire project, but there are some cases where you might wish to vary the calendar by type of resource. The calendar identifies work hours and days. Should you develop all of this for each resource? If you are doing a manual plan, the answer is that you probably will not have time. Even if you are using a computer system, then you may not wish to use the plan for cost analysis.

THE TASK DETAIL

Now it is time for the detailed task information. The task may have a longer description. Keep in mind that the more you write, the more effort it is, not only in generation, but also in maintenance. Also, if you think that if you write more, it will be clearer, you may be wrong. If you do write a description, keep it to a sentence or two.

Each task has a status. It has either been completed, started (in process), or is in the future. Status is important because it allows you to work with a group of tasks with the same status. A task also has resources assigned. If the resources are not available, then you have to decide whether the task can be delayed or divided. If it is divided, then part of the work is done at one time and part at another.

Each task should be assigned resources. There can be multiple resources and several types of resources for a single task. You have to decide on the level of detail of resource assignment. Should you write down all the people who are assigned to the task, or should you only note the person accountable for the task? The answer often depends on project size—larger size projects will more likely use the second approach.

A task can also have a priority level. How would this be used? If two tasks were not competing for the same resources, then priority would not come into the matter. If, however, two tasks were competing, then the task with the higher priority would dominate.

In terms of time and effort, a task can have an elapsed time or duration within which it is to be completed. However, a task also has a specific level of effort as a requirement. For example, task 100 may require three days and an elapsed time of 10 days. In estimating a task effort, three common estimates can be used: optimistic completion, pessimistic completion, and the most likely. Some people create a mathematical estimate using a formula based on these three terms.

Task length and duration get us into scheduling. A task in and of itself can be scheduled in several ways. First, it can be fixed in that the start and end time are locked into place. This is a problem because it removes flexibility from scheduling. We might, however, be forced to fix milestone dates and then work backward using the tasks to establish a schedule.

A second alternative is to schedule the task as soon as possible (ASAP). This means that the task will be scheduled as soon as resources are available and after any predecessor tasks on which the task depends are completed.

A third alternative is to wait until the last minute to do the task. Called as late as possible (ALAP), this approach is applicable to situations where you want to wait until you have received all possible inputs before you do it. It is also used in scheduling to determine how late tasks can be delayed without delaying the overall completion of the project.

SUMMARY TASKS

A summary task contains detailed tasks underneath it. Create the summary or general tasks for a project from the template. The summary tasks assure that the work is of appropriate scope. In our example, the summary tasks are:

1100 Prepare the soil.
1200 Select and purchase the flowers.
1300 Plant the flowers.

We can now expand each summary task into detailed tasks. Additional tasks can be inserted for unanticipated work and rework. This is an alternative preferable to just adding on to the duration of the task. It also helps you to remember what happened in the project several months later. To avoid micromanaging, the project leader can concentrate on summary tasks while the person responsible for a specific summary task can define and work with the underlying detail.

HOW DO TASKS RELATE TO EACH OTHER?

Tasks can be related in several ways. Task A can precede B (sometimes called a series relationship). The two tasks can have no relationships so that they can be done in parallel. Another relationship is that B can be started X days after A starts or started Y days before A ends. This is sometimes called a partial dependency.

These are dependencies based on the work; there are also dependencies because both tasks employ or compete for the same resources. If, for example, two tasks require full-time attention by the same person for one week, then either one task will have to proceed another, or both can be done, but stretched out in time. This is a scheduling issue.

If we have a series of tasks that relate to each other, then we could construct a network or program evaluation and review technique (PERT) chart. Figure 5.1 gives an example of a PERT chart along with some explanatory comments. Most project management software packages include PERT as a way of showing relationships.

In general, we seek to minimize dependencies between tasks. The more dependencies we have, the more complex the schedule is to manipulate and analyze. Large projects have many dependencies so that we seek to create separate but related subprojects. We also try not to have dependencies between tasks at the same outline level. It can be awkward having summary tasks rely on earlier, very detailed tasks.

HOW DO WE PRESENT AND USE THE INFORMATION?

Often, the most popular and most usable way to present project information is to use a GANTT chart. A GANTT chart is a project chart that shows tasks and schedules. It may show task durations and resources. Figure 5.2 presents an example of a GANTT chart and the symbols used in the chart

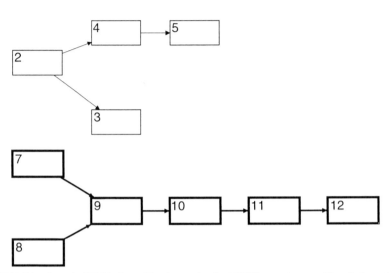

FIGURE 5.1   Sample PERT chart. The boxes in the PERT can give an abbreviation of the tasks and milestones. They can also give schedule information. A PERT chart can be made overly complex if all dependencies involving detailed tasks are shown. For this reason, dependencies between major task areas are sometimes only shown. Note that there is no timescale in the PERT chart. It shows the relationships between tasks independent of time. Also note that in this example, we have split the tasks into three groups. One is overall (tasks 1, 6, and 13). The other two pertain to tasks under 1 and 6. We note that those under task 6 take longer than those under 1 so that they are critical.

for a simple example. Note that the description of the project is at the top of the chart. The left-hand side gives a list of tasks. Other information is given in successive columns including whether the tasks are critical (discussed later in this chapter), the resources needed, and the schedule. In the schedule there are a number of symbols being used. There are different symbols for tasks done, tasks in progress, and future tasks. Note that milestones are shown by the diamond. Due to limited space, standard abbreviations should be used in task titles (see Table 5.2).

The GANTT chart is useful to show status and provide a static view of the project at one time, but it has limitations. As seen, except for the numbering system for the tasks, there is no discernable way to determine how tasks interrelate. This is one reason why it is useful to have both PERT and GANTT charts. We tend to use the GANTT more frequently. Almost

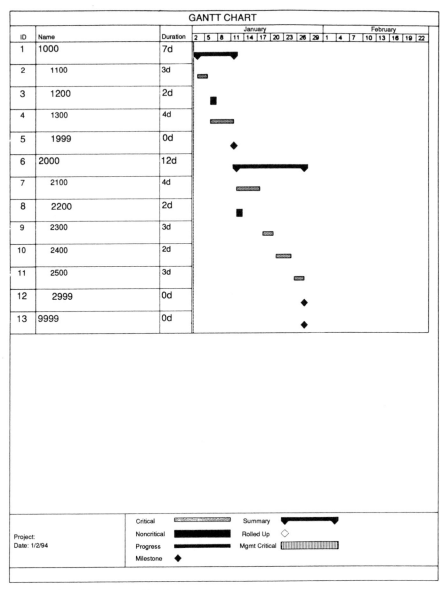

FIGURE 5.2 GANTT chart. The GANTT chart here shows the tasks along with the schedule. The size of the task number indicates its level. Summary tasks are shown in the dark bar. Critical tasks are shown in the dark rectangles. Noncritical tasks are shown in hatching. Milestones are shown as diamonds and have numbers ending in 9.

TABLE 5.2   Examples of Standard Abbreviations[a]

Anal	Analyze
Appr	Approve
Ass (mt)	Assess (Assessment)
Bldg	Building
Comm	Communications
Cond	Conduct
Const	Construct
Coord	Coordinate
Des	Design
Detr	Determine
Dev	Develop
Doc	Document
Eval	Evaluate
Fac	Facility
Genl	General
Hw	Hardware
Impl	Implement
Mgmt	Management
Plan	Plan/planning
Pres	Present
Proj	Project
Remv	Remove
Retn	Retain
Retr	Return
Rev	Review
Sel	Select
Spec	Specify
Strat	Strategic
Sw	Software
Sys	System
Tact	Tactical
WBS	Work breakdown structure

[a] Note that these can be verbs or nouns depending on how the term is used in the task or milestone.

any project management software contains a GANTT chart and most use the chart as the basic means of entering and displaying project information.

SCHEDULING

Let's pause for a moment and see what we have so far. We have created a task structure that we then defined in more detail and interrelated tasks. We then were able to create graphs. This gives us a static schedule. We can now use this as a basis for planning and scheduling.

In planning, we have typically three or more versions of the project plan active at one time. First we have the original signed off and approved plan. This is sometimes called a baseline plan. Second, we have the actual plan, which reflects the actual work done. Third, we have the future scheduled plan. Finally, we could have several optional contingency and projected plans. Each of these plans would use the same tasks, but some of the details and interrelationships might be different.

Here are some examples of potential "what if," analytical schedules we might want to consider:

- All future tasks are as soon as possible. This provides us with a schedule that shows us how we could finish the project at the earliest time.
- All future tasks are scheduled as late as possible. This shows how much work can be delayed without causing the schedule to slip.
- Schedule subset. As opposed to the entire schedule, we could extract a set of important tasks and milestones. An example here is a summary schedule. In some software systems, the summary schedule can be obtained from a roll-up of the detailed tasks.

What we do first is to develop a basic schedule for which we seek approval. As work is performed, we update the schedule. This moves tasks from future to being started to being complete. This is our actual schedule. We can develop our projected schedules by changing future completions of tasks and task effort and then comparing these to the planned and actual schedules.

What is the scheduling process? After setting up the schedule, we first determine the managerial and mathematical critical path. This is discussed in detail in the next section. The critical path is the set of tasks and milestones such that if delays occur on the critical path, the project is delayed. We want to focus on this because of its importance. If the scheduled completion of either parts or all of the work are not acceptable, then we can work with those parts as opposed to the entire schedule.

What other changes can we make in scheduling? Here are some possibilities:

1. We can change the details of specific tasks in terms of effort, duration, and so forth. This is a detailed level change.
2. We can alter the dependencies between tasks. This change is at a higher level.
3. We could change the allocation of resources. This is called resource leveling, which we discuss later.

After you do the changes, you then have to compute an overall schedule again. With a large schedule this can be very complex and tedious if done manually. That is another reason for using project management software. We will return to scheduling after we discuss the critical path.

THE CRITICAL PATH: MYTH OR REALITY

We explained that the critical path is the path from the start to the end of the project such that if any task is delayed in the path, the project is delayed. This is the technical, mathematical definition of critical path. When you first encounter this definition, it makes sense. A path in the project has to go from beginning to end. The critical path is the longest path in the project. Any other tasks not on this path have positive slack time. Slack time is the time lag necessary to move a noncritical task into the critical path. For example, if we say a task has a slack of three days, then if its schedule slips three days, the task moves to the mathematical critical path. It follows that critical path tasks have zero slack time. The algorithm to develop of mathematical critical path is called the Critical Path Method (CPM).

The definition is fine for many projects. However, it can happen that some tasks not on the critical path are very important. There can also be simple, short tasks that mathematically happen to fall on the critical path. The presence of both or either of these conditions leads to the questioning of the validity of the schedule and the critical path.

These considerations give rise to creating a separate critical path. This path we call the managerial critical path. It may or may not include any or all of the tasks on the mathematical critical path. Figure 5.3 gives an example of a project with separate managerial and technical critical paths. Constructing the managerial critical path is a matter of judgment and assessment of the project plan. If we have several high risk tasks, we will have a number of management critical paths.

The reason we call this section myth or reality is that the myth may be the mathematical critical path, and the reality is the managerial critical path. Many complex projects might start with the mathematical critical path. Then the managerial critical path is constructed from the mathematical critical path by first removing any small, less important tasks from the path and then inserting any important tasks that are on the path.

Note that if we change task detail, dependencies, or other project information, then the critical path may change as well. If we change the method of scheduling tasks, we can dramatically alter the critical path. This happens often in resource leveling (the process of moving tasks around to accommodate limited resources).

There are some other definitions we can make as a result of defining the critical path. If a task is not on the mathematical critical path, then we could ask how much time would be needed in terms of slippage before the task becomes critical. More precisely, we define slack time as the difference between the latest time that the task can be started (without delaying the project) and the earliest time it can be started.

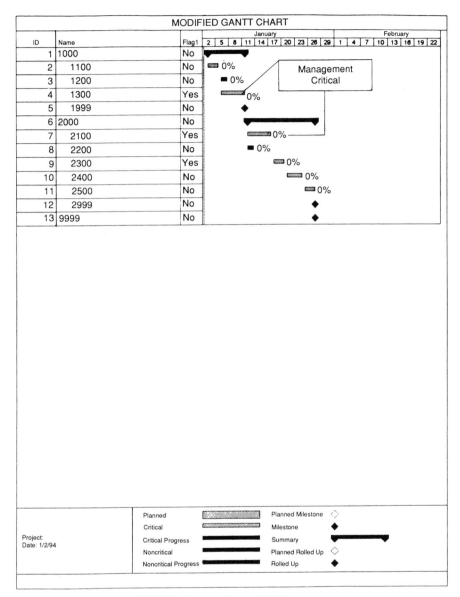

FIGURE 5.3 Modified GANTT chart.

In project management, we use the critical path as a powerful political tool. In a meeting we can use the information to impress on specific project team members the importance of their tasks and why they are critical.

Showing the critical path adds credibility to the argument. Using the management critical path can be even more powerful.

RESOURCE CONFLICTS AND LEVELING

Resource conflicts occur when we allocate more of a resource than what is available. This is a problem related to time. We should be concerned about overallocation in the next month, more than five months from now when the situation could change. We can also have overallocation among projects in cases where several projects share the same resources. A resource can be overallocated. That is, you require more of a particular resource than is available. What can you do within the current project structure to resolve this? Resource leveling. You can move tasks to other time periods with lower resource demands. You can delay tasks that are concurrent. In real life you could allow for overtime, change the calendar, change the plan, assign other resources to mention a few methods.

How do you do resource leveling? Obviously, you make changes to the schedule and then see the results. Software can help you do this. Again, keep in mind that employing project management software to do resource leveling can substantially change the schedule. If you do it manually or automated, you first must look at specific resources that are a problem. Then you look at those tasks to which the resources have been applied.

THE CONCEPT OF A FILTER

We have often asked you to extract from the schedule or plan certain tasks as critical or specific resources for analysis. This is a very important part of project management analysis. In project management the term used is filter. Applying a filter means extracting a subset of the schedule using specific criteria. Criteria might include dependencies, resources, time periods where tasks were to be done, late tasks, and so forth. An example is to extract all tasks from the schedule with a specific level of risk. Filtering is another argument for automated tools in project management because this can be used easily and quickly. Moreover, when you develop a filter, all of the tasks, dependencies, resources, and schedule are preserved in the extracted schedule.

WHAT IF YOU DON'T LIKE THE SCHEDULE?

Just because you spent a lot of time and effort on developing the schedule and it does not work out, you should not be discouraged. Put down the schedule and do something else for some time. When you go back to it, proceed backward. Ask yourself what you don't like about the schedule. Here are some symptoms of problems with a schedule:

- The milestones are too late.
- Too much work is being done in one period.
- Some resources are overcommitted.

We can identify some specific steps, amplifying on those mentioned earlier:

- Add more resources.
- Break up larger tasks into smaller ones.
- Increase parallel effort.
- Substitute tasks with others.
- Eliminate tasks.

DEALING WITH RISK

How do you deal with risk in the project? One way is to be generous in your project task estimates of effort. Another way is to consider the schedule overall. We suggest that you get the uncertainty identified early in the project. If there are specific unknowns, then you could create milestones that refer to the specific risks. In a software project, you could have a milestone that the software operates within the capacity of the machine. In a physical project, you might have obtaining permits, doing soil surveys, and so forth as milestones. In an engineering project, you might have the creation of a prototype and its testing as a major milestone. Whatever the project, you want to get the risk identified out front. If you have specific unresolved issues that persist in the project, then when they are finally resolved, they may destroy or delay the project.

Beyond this you should have the specific areas of project risk reflected in the tasks and the milestones. That is, if you have an issue or problem that has to be resolved, then it should refer to specific tasks and milestones. Another idea is to create a resource called risk and then apply it to specific tasks and milestones. When you want to find the areas of risk, apply a filter to the project based on the resource "risk."

ADMINISTRATION VERSUS MANAGEMENT

Setting up a project sounded at first like an administrative task. But developing the schedule is one of the most creative parts of the schedule. If you view developing a schedule as just creating a list, then you are focusing on administration. If you focus on how you can get the most of the resources within the schedule and are using the schedule structure in a political sense, then you are adopting a management approach.

EVALUATING FOR FLEXIBILITY AND MAINTAINABILITY

Our analysis of a plan has focused on costs, time, and resource use. These are basic parts of the scheduling process. But you will be using the schedule and plan as a living, breathing entity. You will have to update it; you will have to change it. What do you do? You should evaluate the schedule in terms of flexibility and maintainability. Make a list of events that could occur and see what impact they have on your schedule. Some examples are as follows:

1. What if the project is scaled back dramatically?
2. What if management suddenly wants work speeded up?
3. What if a dependent project is killed or fails. What will be the impact?
4. What if critical resources are removed from the project?
5. What if the project manager is replaced?

How do you answer the questions? Create a new version of the schedule for each of the "what if" questions. This will tell a lot about the impact of changes on the project.

WHAT YOU HAVE WHEN YOU ARE DONE

After doing all of this work, you have a schedule that has been developed through analysis. Why do all of this work? Why not just get started and then revise it later? There are several reasons why this is not a good idea:

1. In a presentation you will be asked what happens if some change is made. You can then respond based on analysis and data.
2. If management wants the project to be done earlier or with fewer resources, you either have already done this analysis, or you can easily do it based on your experience.
3. Doing the analysis makes you the expert on this particular schedule. You will be judged as a project manager by how well you are doing.

USING A NETWORK AND THE WEB

In a shared network individuals on the team can supply details for their tasks to the same project file. We can employ electronic mail, groupware, and the Web to initiate and support a discussion among the team.

Couldn't you just get people in a room and do this? Perhaps, but using the network tools provide for convenience, less intimidation of junior staff, and interaction. As people get in the habit of working with shared information, consensus is often easier to build.

A LANDSCAPING EXAMPLE

It is time to consider an example that will tie these pieces together. Figure 5.4 gives a task list for a landscaping project at home. We could

LANDSCAPING - TASKS, PREDECESSORS, RESOURCES

ID	Name	Predecessors	Resource Names
1	1000 Review Current Yard		
2	1100 Remove Debris		FAMILY
3	1200 Clear Weeds, Dead Trees		FAMILY
4	1300 Remove Extra Soil/Cement		FAMILY
5	1999 M: Basic Yard Cleared	2,3,4	MILESTONE
6	2000 Develop Plan	1	
7	2100 Select Architect		
8	2200 Review Yard	7	ARCHITECT
9	2300 Develop Plan	8	ARCHITECT
10	2400 Present Plan	9	ARCHITECT
11	2500 Review Plan	10	FAMILY
12	2600 Revise Plan	11	ARCHITECT
13	2999 M: Plan Approved	12	MILESTONE
14	3000 Install Watering System	6	
15	3100 Select Plumber		YOU
16	3200 Develop Plumbing Plan	15	PLUMBER
17	3300 Review Plan	16	ARCHITECT,YOU
18	3400 Do Plumbing	17	PLUMBER
19	3500 Review Work	18	ARCHITECT,YOU
20	3999 M: Plumbing Complete		MILESTONE
21	4000 Implement Rock Work	6	
22	4100 Visit Stores		ARCHITECT,FAMIL
23	4200 Select Rocks	22	ARCHITECT,FAMIL
24	4300 Purchase Rocks	23	YOU
25	4400 Deliver Rocks	24	STORE
26	4500 Install Rocks	25	WORKMEN
27	4999 M: Rocks Installed	26	MILESTONE
28	5000 Get/Plant Trees, Shrubs	6	
29	5100 Trees		
30	5110 Evaluate/Select Trees		ARCHITECT,FAMIL
31	5120 Order/Purchase Trees	30	FAMILY
32	5130 Prepare Soil for Trees	31	FAMILY
33	5140 Take Tree Delivery	32	STORE
34	5150 Plant Trees	33	WORKMEN
35	5199 M: Trees Planted	34	MILESTONE
36	5200 Shrubs		
37	5210 Evaluate Shrubs		ARCHITECT,FAMIL
38	5220 Buy Shrubs	37	YOU
39	5230 Store Shrubs	38	FAMILY
40	5240 Plant Shrubs	39	FAMILY
41	5299 M: Shrubs Planted	40	MILESTONE
42			
43	6000 Plant Flowers	39,14,21	
44	6100 Prepare the Soil		FAMILY
45	6110 Get Tools		YOU
46	6120 Dig up Soil/Remove Weeds	45	FAMILY
47	6130 Fertilize Soil	46	FAMILY
48	6199 M: Ground Ready for Flower	47	MILESTONE
49	6200 Select/Purchase Flowers	44	
50	6210 Select Flowers		FAMILY
51	6220 Identify Fertilizer		YOU
52	6230 Purchase Flowers/Fertilizer	50,51	YOU
53	6240 Take it Home	52	YOU
54	6299 M: Flowers Ready for Planti	53	MILESTONE
55	6300 Plant Flowers	49	
56	6310 Prepare Soil		FAMILY
57	6320 Plant Flowers	56	FAMILY
58	6330 Water Flowers	57	FAMILY
59	6399 M: Flowers are Planted	58	MILESTONE
60	9999 M: Garden Complete	43	MILESTONE

FIGURE 5.4 Tasks, predecessors, resources chart. This simple chart is for landscaping a garden with a landscape architect. There are two resources: the architect and you.

have picked some arcane business example, but we are trying to focus on methods, not on some specific type of project. Note that the project includes plumbing, electrical, trees, plants, and cement and bricks. Note that the milestones are based on what individual workers will be doing.

The PERT chart for these tasks is given in Figure 5.5. Note that in the early stages, a lot of the work can be done in parallel. In later stages, it becomes sequential because you are working in a yard around what has already been done. Note also that some of the work such as lighting is done in several parts. Later parts depend on the planting being in because the lighting will have to be adjusted for visual appeal.

The resources are the architect, yourself, a plumber, and electrician. If we develop estimates of tasks and apply the resources, then we would get the GANTT chart in Figure 5.6. Note that we could filter this plan on milestones and get Figure 5.7. This smaller list can keep your sanity as your yard becomes torn up. Now let's suppose that you don't like this schedule because it stretches out for five months. First, we can see if some of the tasks can be moved up for parallel effort. If we make the design and clearing in parallel, we can speed up the project as in Figure 5.8. More coordination is needed. Better add more coordination tasks and some reviews to ensure that work doesn't have to be redone. The result is Figure 5.9.

EXERCISES

What should you do now? First, you need to gain experience in the setup of a project plan. You should do several plans to get practice in defining tasks, milestones, and so on. One of the projects you develop should have at least 50–75 tasks with parallel and dependent tasks. Call the results of this part the baseline schedules.

The second part is to develop the critical path. You should think about the project and develop the mathematical critical path. Then, using your ideas, create the managerial critical path. Compare the two.

The third part is more fun. This is the analysis of the schedule along the lines we have suggested. You would do a "what if" analysis for different alternatives. You should try most of the following:

- Changing dependencies
- Adding resources
- Doing resource leveling

Obviously, this effort will be easier if you are using software. However, learning the software is an effort unto itself. We recommend that you do the first schedule manually. Then you can later use a software package. You will appreciate the automation and tools and know better how they can be used.

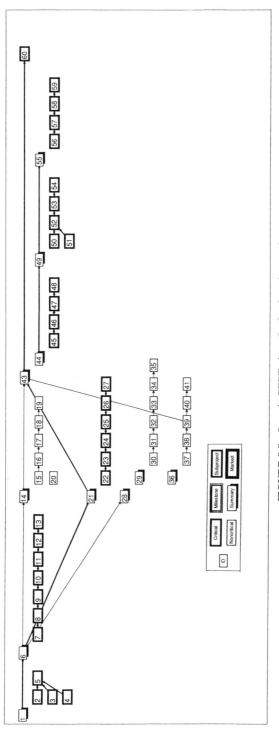

FIGURE 5.5 Sample PERT chart for landscaping project.

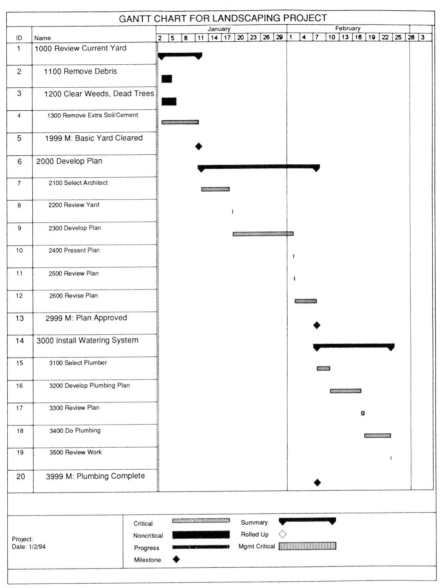

FIGURE 5.6   GANTT chart for landscaping project. The GANTT chart here shows the tasks along with the schedule. Critical tasks, milestones, and other tasks are shown with different symbols. All of the tasks are performed by you except those performed by the architect.

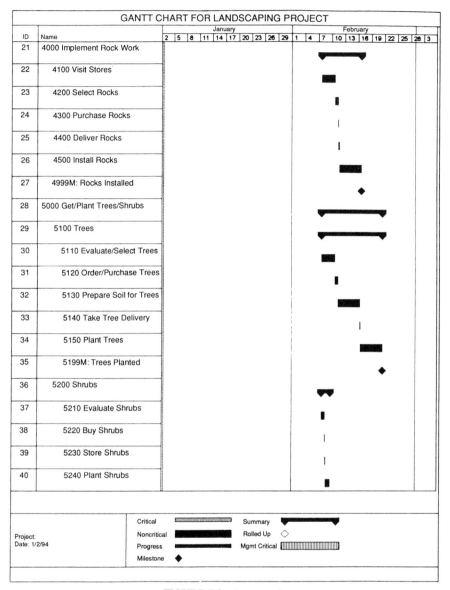

FIGURE 5.6 (*continued*)

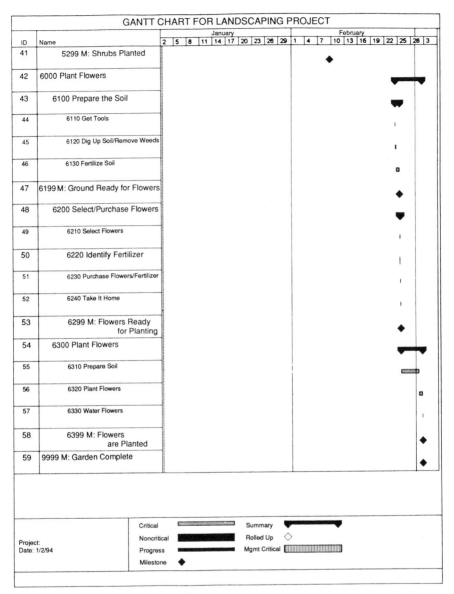

FIGURE 5.6   (*continued*)

MILESTONES FOR LANDSCAPING PROJECT

ID	Name
5	1999 M: Basic Yard Cleared
13	2999 M: Plan Approved
20	3999 M: Plumbing Complete
27	4999 M: Rocks Installed
35	5199 M: Trees Planted
41	5299 M: Shrubs Planted
47	6199 M: Ground Ready for Flowers
53	6299 M: Flowers Ready for Planting
58	6399 M: Flowers are Planted
59	9999 M: Garden Complete

FIGURE 5.7 Milestones for landscaping project.

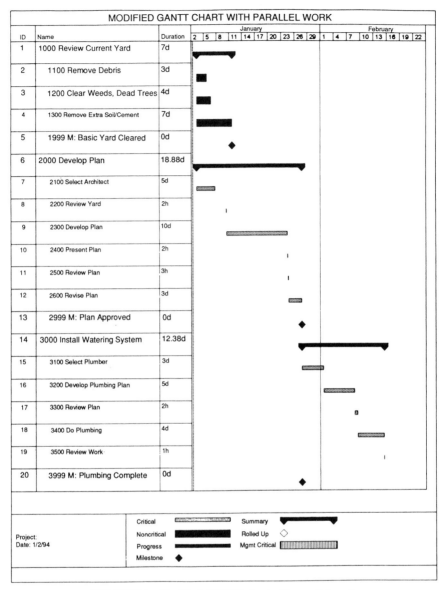

FIGURE 5.8   Modified GANTT chart with parallel work.

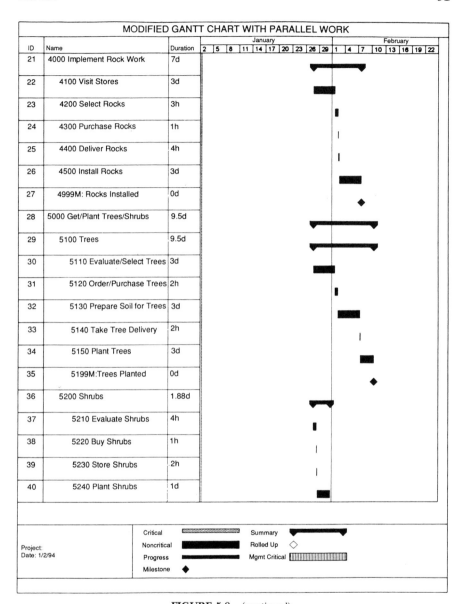

FIGURE 5.8 (*continued*)

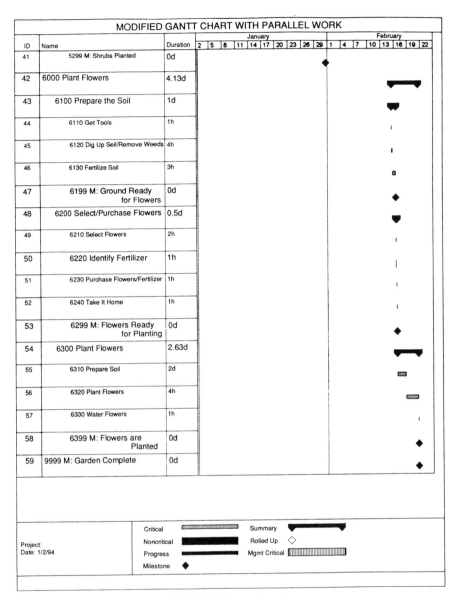

FIGURE 5.8   (*continued*)

MODIFIED PLAN WITH SEVERAL COORDINATION TASKS

ID	Name	Predecessors	Resource Names
1	1000 Review Current Yard		
2	1100 Remove Debris		FAMILY
3	1200 Clear Weeds, Dead Trees		FAMILY
4	1300 Remove Extra Soil/Cement		FAMILY
5	1999 M: Basic Yard Cleared	2,3,4	MILESTONE
6	2000 Develop Plan		
7	2100 Select Architect		
8	2200 Review Yard	7	ARCHITECT
9	2300 Develop Plan	8	ARCHITECT
10	2350 Interim Review		
11	2400 Present Plan	10	ARCHITECT
12	2500 Review Plan	11	FAMILY
13	2600 Revise Plan	12	ARCHITECT
14	2999 M: Plan Approved	13	MILESTONE
15	3000 Install Watering System	6	
16	3100 Select Plumber		YOU
17	3200 Develop Plumbing Plan	16	PLUMBER
18	3300 Review Plan	17	ARCHITECT,YOU
19	3400 Do Plumbing	18	PLUMBER
20	3450 Review Work as it is done		YOU
21	3500 Review Work	20	ARCHITECT,YOU
22	3999 M: Plumbing Complete		MILESTONE
23	4000 Implement Rock Work	6	
24	4100 Visit Stores		ARCHITECT,FAMIL
25	4200 Select Rocks	24	ARCHITECT,FAMIL
26	4300 Purchase Rocks	25	YOU
27	4400 Deliver Rocks	26	STORE
28	4500 Install Rocks	27	WORKMEN
29	4550 Review Work		YOU
30	4999 M: Rocks Installed	29	MILESTONE
31	5000 Get/Plant Trees, Shrubs	6	
32	5100 Trees		
33	5110 Evaluate/Select Trees		ARCHITECT,FAMIL
34	5120 Order/Purchase Trees	33	FAMILY
35	5130 Prepare Soil for Trees	34	FAMILY
36	5140 Take Tree Delivery	35	STORE
37	5150 Plant Trees	36	WORKMEN
38	5155 Review Work		YOU
39	5199 M: Trees Planted	38	MILESTONE
40	5200 Shrubs		
41	5210 Evaluate Shrubs		ARCHITECT,FAMIL
42	5220 Buy Shrubs	41	YOU
43	5230 Store Shrubs	42	FAMILY
44	5240 Plant Shrubs	43	FAMILY
45	5299 M: Shrubs Planted	44	MILESTONE
46	6000 Plant Flowers	43,15,23	
47	6100 Prepare the Soil		FAMILY
48	6110 Get Tools		YOU
49	6120 Dig up Soil/Remove Weeds	48	FAMILY
50	6130 Fertilize Soil	49	FAMILY
51	6199 M: Ground Ready for Flower	50	MILESTONE
52	6200 Select/Purchase Flowers	47	
53	6210 Select Flowers		FAMILY
54	6220 Identify Fertilizer		YOU
55	6230 Purchase Flowers/Fertilizer	53,54	YOU
56	6240 Take it Home	55	YOU
57	6299 M: Flowers Ready for Planti	56	MILESTONE
58	6300 Plant Flowers	52	
59	6310 Prepare Soil		FAMILY
60	6320 Plant Flowers	59	FAMILY
61	6330 Water Flowers	60	FAMILY
62	6399 M: Flowers are Planted	61	MILESTONE
63	9999 M: Garden Complete	46	MILESTONE

FIGURE 5.9 Modified plan with several coordination tasks.

GENERAL EXAMPLE:
MANUFACTURING ORGANIZATION

Recall our manufacturing example from the previous chapter. We had selected integration and final assembly as the first project. If the first project is to serve as a model for later projects under our strategy, we must develop a generic planning approach and method for setting up the project that can be adapted for later use. In this case, a standardized set of tasks and milestones was defined along with a list of resources. A set of assumptions was developed for setting up the plan. The overall plan was divided into areas. Each area was under the direction of a specific manager. Within an area, 3–5 critical processes were defined. If these processes were completed, then the area was completed in terms of work. Under each process, many tasks were defined. The area and process levels of the project plan are summaries of the detailed tasks and visible to all. The process and task details were the responsibility of the area manager.

In terms of defining tasks, the attention was toward ensuring that a task was associated with physical work that could be reported on in the normal course of a day. A task can be supported by a checklist. If the items in the checklist are completed, the task is complete. This required software tools that supported checklists linked to the project schedule. Milestones included not only the completion of major work activities, but also the receipt and hand-off of manufacturing components. This would allow management to be aware of when schedules slipped due to late delivery.

Resources included about 50–60 key elements including equipment, facilities, and personnel types. There are over 1000 individual resources. Managing at that level would have been impossible, so the decision was made to focus on only the critical resources. Given the variability of the manufacturing process, the standard critical path held little meaning. The management critical path identified tasks with risk, and attention was focused on these. These tasks were not difficult to identify given the substantial history of the previous manufacturing.

As we have discussed, there are summary level tasks that represent a group of lower level tasks. One can have several different levels of tasks. A rule that was established was that dependencies between tasks were only allowed at the same level. Thus, summary level tasks depend on each other, but the subtasks under a summary level task only depend on other tasks within the same summary level task. The benefits of this are (1) it is easier to view and change because dependencies are grouped; and (2) when extracting a group of tasks, one can extract or filter a group under a summary task and the surrounding summary tasks. The second benefit reduces the amount of data that is being viewed and changed.

This approach provided accountability for the work as well as visibility of project information to other area managers so as to aid their planning efforts. This approach also provided flexibility in that if the product changed, the tasks could change, but the area and summary processes remained the same—giving stability and supporting analysis across multiple products.

Let's first review the process. We first constructed a schedule bottom up. That is, we defined a series of tasks and relationships and so forth and created an initial plan. The second general step is to do scheduling. In scheduling we not only try to improve a schedule, but we also attempt to better understand the information. You see the schedule is a collection of small pieces of data. You do not know how it will come out until you put it all together. That makes it different from other types of activities. By doing scheduling and by trying out different alternative schedules, you can get a sense for where the bottlenecks as well as points of flexibility are.

A primary goal in doing scheduling is to obtain a schedule which satisfies

- Lowest cost
- Smallest elapsed time
- Minimal risk

Other goals are

- Flexibility—Can the schedule accommodate major changes?
- Maintainability—Can the schedule accommodate minor changes?
- Accountability—Are the tasks defined in such a way as to provide accountability?

There are trade-offs here. Very seldom do you get a schedule that meets all of the above criteria and goals. Even when it does future changes and events may affect the degree to which you have achieved these goals.

6 Being a Winning Project Manager

Developing the project plan provided insight into the thinking process. Now we can study the people side of projects. Substantial projects require more than one person to succeed. This group of people is commonly referred to as the project team. Any team of people must be managed; hence, the need for a project manager.

What does a project manager do and how does work get done? Are there some guidelines to prevent failure and achieve success? We consider becoming a project manager as well as avoiding being named as one. This last point is not as strange as it sounds. Good project managers are hard to find. When an organization finds one, it is natural to move the project manager from one project to another. This works for a while, but gets tiresome after the fourth or fifth major project. The problem then is how to escape the role without risking your job security.

WHAT DOES A PROJECT MANAGER DO?

We have already seen some of the duties of a project manager in terms of managing the team, setting up the schedule, and controlling the project. These are only a few of the most obvious duties. Table 6.1 presents a list of potential duties. These items can be considered as either managerial or administrative. Our primary concern is management, but we briefly discuss administration.

A project manager has three clearly defined management responsibilities:

1. Resource manager. Manage and direct project resources to achieve the project objective.

TABLE 6.1   Examples of Duties of Project Manager

Define the project.
Prepare the project plan.
Present and sell the project and project plan.
Convey the purpose of the project.
Implement organization policy and procedures.
Apply effective project management methods and tools.
Interact with general management.
Interact with line managers.
Obtain project resources.
Coordinate the use of project resources.
Interact with customers and suppliers.
Interface with other projects.
Prepare and review budgets.
Review milestones.
Develop schedules.
Revise schedules.
Revise budgets.
Manage the project team.
Delegate responsibilities.
Define issues, problems, and opportunities.
Participate in specific project tasks.
Analyze issues.
Prepare management presentations.
Make management presentations.
Determine project status.
Perform "what if" and projection analysis for the project.
Deal with conflict with management.
Deal with conflict within the project team.
Restructure project plan.
Understand the tools and methods used in the project.
Learn from previous projects.
Deal with the impacts of project change.
Coordinate the removal of project team members.
Coordinate the arrival of new project team members.

2. Planning and control manager. Develop the project plan and ensure that the work is completed on time, within budget, and with acceptable quality.
3. Coordinator. Interface with upper management regarding project review, approval, and address project issues. The manager must also relate successfully to line managers and staff.

Often, attention is focused on resource management. In reality, though, a manager may spend much more time performing the other two roles. In ancient Rome, a Roman governor who was sent to a remote province had specific responsibilities in the province, but all successful managers including Caesar and Augustus paid attention to and looked behind their back at Rome. Being a success in the province might not matter if you ran into political problems in Rome. With travel and distance barriers, how could this happen? All provinces had to supply Rome with men for the army or navy, and food, manufactured goods, and other materials. You could be popular in the province, but if you ran afoul of Roman politics, you failed. Similar examples in the United States in military history abound. We have successes in Washington, Lee, Grant, Pershing, and Eisenhower. We have failures in early generals in the Union Army during the Civil War.

This brings us to an observation of successful project management. As a project manager, you must identify and satisfy your audience. If you think that the audience is the project team and the eventual users of the end product of the project, you are deluding yourself. We suggest that you assume that the audience includes all of those related to the project, general management, and line managers. Following up on this idea, you can see that planning and control and interface management are very important.

Look at a project manager role now from a perspective of doing things. Here is another list:

- Directing. Direct the project resources.
- Motivating. Motivate the project team in the presence of normal pressures of work as well as political realities and pressures.
- Planning. Anticipate and plan ahead. Anything unforeseen can reflect poorly on the project manager, even if it is not the manager's fault.
- Supervising. Supervise the work.
- Administering. Today's project manager often has to perform many administrative tasks without support.
- Interfacing and coordinating. This is the set of tasks that involves meetings and working with line management, external contractors, subcontractors, and management.
- Doing. Do some of the tasks themselves and do not just manage the project.
- Training. Train members of the project team in project management methods and tools in which they are involved.

- Counseling. Act as counsel to management on specific technical or business issues as well as to staff on project and even personal issues.
- Delegating. Delegate and then supervise work on the project.
- Resolving conflicts over resources and schedules.

Another reality of project management is that there are no excuses. The project is visible. If there is failure or problems that are not solved, then the organization may seek to find a scapegoat. This seems harsh, but you should make this assumption so that you will be more alert to potential problems. As project managers we have found that if we make these reasonable political assumptions, we will be able to solve issues earlier and at lower organizational levels.

THREATS TO PROJECT MANAGERS

What is waiting in the wings? There are people who may be jealous of the resources and power of the project. They see that their own staff or projects are drained away or put on hold because of your project. They may pay lip service to supporting your project, but underneath they are seething and resentful. Some managers may have candidates for managing the project if they can just get rid of you. They keep looking for ways to involve their candidate in the project. Then with knowledge of the project, they can watch for a false step or problem, and, then like a leopard, they can pounce.

Another threat is that some people may feel threatened by the project itself. If the project succeeds, their department, role, and even their jobs may be in jeopardy. Today and in the future, this will be true because many projects will act to transform organizations to gain efficiency and effectiveness through reengineering.

This sounds very negative and depressing, but it should not be taken that way because it is a fact of life. What should you do as a project manager? Do you attempt to deal with these factors? You have to assume that they are always present. You should always assume that people will act out of their own and their organization's self-interest.

BASES OF POWER FOR A PROJECT MANAGER

A project manager is the manager of the project. As such, the person has power. There are several different aspects to the power of the project manager that can be invoked in positive or negative ways.

- Formal authority. This is the case when the project manager uses his or her formal power to make decisions and decide issues. This may have to be used occasionally, but it can easily be abused. You can successfully appeal to authority very few times.

- Reward and penalty power. This is the power to reward or punish team members. In project management, the project manager can bestow few tangible rewards because the manager is not the normal supervisor of the team members. The same applies for penalties. We have found that this type of power must also be used sparingly.
- Power as a technical authority. The project manager is an expert and so the team members respect the manager because of the expert knowledge. This is rare in many projects. However, with experience, a project manager can become known as an expert in managing projects.
- Power through respect. Through the experience and behavior of the project manager, the project manager instills a sense of respect among the project team. This is probably the best type of power. It is the most honest and direct, and will tend to be long lasting.

TYPES OF PROJECT MANAGERS

Over the years we have been able to distinguish a number of different types of project managers.

- The administrator. This is the project manager who is engrossed in administrative trivia. This person typically produces very good reports and presentations, but, unfortunately, the work may not get done.
- The doer. This is the person who takes over tasks and does the work. This person does not delegate well. The manager typically micro-manages the staff on their tasks. Not a healthy situation.
- The upward-oriented manager. This manager focuses attention on management and getting visibility for themselves with upper management. They may not care that much for the project and may even see it as a stepping stone to something better. They tend to delegate the project. The team tends to see through this and resents the manager.
- The task master. This project manager will push project team members as long and as hard as possible to get the work done. While this might be noteworthy, the person typically does this when there is no pressure.
- The leader. The person supports the project team members to get the work done and focuses on issues and not administrative trivia.

ATTRIBUTES OF A GOOD PROJECT MANAGER

We have already seen that the good project manager should develop respect among team members and management as a source of power.

COMMUNICATOR

A project manager needs to communicate well verbally and in writing within and outside of the project. Communications problems have been identified as a major source of project failure.

GENERALIST

The project manager has been able to see the big picture and then relate it to the current project situation. The manager must then convey the meaning of this to management and the team. What is the big picture? It includes the relationship of the project to the company. It is the potential impact of the end product of the project to the firm.

PROBLEM AND CONFLICT SOLVER

Of all of the attributes of project managers, this is the one that has remained important for centuries. A project manager must be able to identify and understand problems, place them in perspective, and then develop and implement solutions. Being able to identify, analyze, and resolve issues are critical success factors for project leaders.

PEOPLE MANAGEMENT

The project manager has to be able to effectively manage people. A manager can be weaker here if the manager has other compelling strengths. But there is not much room for weakness. We will talk more about this below when we discuss motivation.

EXPERIENCE

It is not enough to have had experience in several projects. You have to be able to take this experience, integrate it, and then apply it to the current project. Otherwise, there is no learning or wisdom gained.

AMBITION

It is appropriate that a project manager have ambition. Otherwise, why be a project manager? The positive side of ambition will lead you to work hard and positively. It will also help you in dealing with people and getting the project completed. In excess, ambition can do all of the things that have been written about in terms of corruption, deceit, and other similar traits. But it is not wrong to be ambitious, but ambition must be tempered.

ENERGY

A project manager has to have a lot of energy. Not just to deal with problems, but to be able to take measured risks. Long hours are a part of the job. In the past people thought that this meant that project managers had to be younger people. Experience and knowledge are valuable.

KNOWLEDGE

A project manager needs to be able to acquire information about all aspects of the project quickly. Thus, it is not just knowledge, but it is also the velocity by which it is acquired and the ability to apply the knowledge in the project.

PERSPECTIVE

A project manager has to be able step back from the project and take an overall view. Otherwise, symptoms of problems may be overlooked.

SENSE OF HUMOR

Events, setbacks, and successes need to be taken seriously, but you must also have a sense of humor.

INITIATIVE AND RISK TAKING

A project manager should be able to show initiative and be willing to take risks. What do we mean here when we have been stressing minimization of risk? Taking risks may mean giving tasks to junior staff who have not done that exact type of work before, or having faith in some things that are not under your control.

BEING ORGANIZED

Because project managers are likely to be playing a dual role—project manager and project administrator—they have to be organized. This includes managing time and being able to allocate time and energy to issues that are important. At times it means taking an overall view. At other times, it means being very focused and getting a specific set of tasks done or following an issue to its conclusion.

ABLE TO TAKE DIRECTION AND SUGGESTIONS

A good project manager has to be able to admit that there is a problem and accept responsibility. They have to be able to accept suggestions, hints, and criticism and turn it into positive action. Ignoring problems can be their undoing and a source of failure. People will not offer criticism if they know that it will not go anywhere.

FAMILIARITY WITH THE ORGANIZATION

This is not just memorizing an organization chart. It means knowing how the organization works. Where are decisions really made? Who really has power? Which managers rely on lower level managers and staff? Who can make decisions? It takes time to learn this, but you have to work at it. How do you learn the organization? The basic method we have found that works is observation. When you are in a meeting, observe how people interact. Observe who makes decisions and who implements. Also, look for the people who have the knowledge of the organization and who seem to have little or no problems getting decisions made and obtaining resources.

KNOWLEDGE OF TECHNOLOGY

If the project is based on specific technologies, the project manager has to gain some basic knowledge of the technology. Otherwise, the manager will lack credibility with the team. The manager may get the reputation of being easily "snowed" or sold on some idea by appealing to the technology. Now we do not mean that the project manager has to be a techie or expert, but you need to have some basic knowledge.

MATURITY

Project managers have to learn to take things in stride. If they lose on an issue, they cannot accept it as fatal. The issue can probably be resurrected later at a more appropriate time. Maturity also means having a sense of timing of when to make certain moves.

TOUGHNESS

At times in a project, project managers are faced by personal and management challenges. It is often easiest to take the path of least resistance and either defer decisions or cave into one position. Being tough means that you are willing to take a contrary position.

Obviously, these skills are interdependent. Experience and knowledge facilitate problem solving. Managing people and communicating go hand in hand. We have identified these separately even though they are closely linked so that you can work on improving your skills in areas where you see the need and benefit from enhancement.

MOTIVATING THE PROJECT TEAM

Specific characteristics and situations in dealing with team members are discussed later. Here we devote our attention to the challenge of motivating the project team. Project managers have to provide motivation and leadership. There are several aspects to motivating people:

- Appealing to the project and its eventual positive impact on the organization. This can be used with the team as a whole. It cannot be overused; otherwise, it will sound trite.
- Appealing to the team member's self-interest. Here you focus on what the individual team member will gain from the project. First, there is the experience of the project. This can be useful if there are other similar projects that will be coming up. Second, you can show that the person is gaining technical skills and competence that will help them wherever they go. Third, by working on the project, they show that they can be part of a team. Today, as we stated in the first chapter, more and more work requires coordination and cooperation. You must provide motivation and encouragement to individuals in the team more than the team as a whole. Doing this helps the project leader keep up-to-date on what is going on in the project.

SETTING UP ELECTRONIC COMMUNICATIONS

The project leader should not only define how electronic mail, the Web, groupware, and network will be used, but also facilitate getting these established with procedures at the start of the project. Waiting until the middle of the project will be too late since people have already developed different communications habits.

CONSTRAINTS ON PROJECT MANAGERS

A project manager has to work with many different constraints. First and, perhaps, most important are political constraints. The project manager has a very limited scope of power. This is reflected in resources because the manager has control over resources for a limited time. It is also reflected in the budget because the project manager has no standard budget as does a line department. The manager has direction over limited funds that are allocated to specific purposes and that are available for only a limited time.

Another set of constraints is the societal environment at the time that the project is undertaken. This includes the attitude toward work and ethics.

Today we have several constraints, including time and a demand to show early results. Patience is somewhat of a luxury. Even large projects that are approved require periodic approval and endorsement. This is based on showing results. Contrast this with the 1940s–1960s. Then there were many projects (atomic bomb, other weapons systems, alternative energy, etc.) where much time and resources were allocated over a longer horizon. Today, the time horizon is compressed.

PROJECT MANAGEMENT VERSUS LINE MANAGEMENT

In Chapter 2 we highlighted the differences and potential conflict between the project and the line or standard organization. This conflict and these differences impact the role of the project manager. Table 6.2 gives a list of typical differences between a project manager and a line manager and indicates why there is a conflict. There are differences in the two types of management in susbtance, roles, focus, and styles required. A project manager is closely related to a manager of a firm who has been hired to turn around a company and then move on to another situation. Project managers focus on getting the project done. They care less about items outside the project. A line manager, on the other hand, deals with standard repeating patterns of work. There may be some small projects in his or her group, but the attention is on day-to-day work.

SETTING UP A PROJECT

We have talked about the role and environment of project management in general. Now it is time to turn to details.

The project manager must first set up the project along the lines discussed in Chapters 4 and 5. Setting up a project involves doing work, gaining involvement by organizations to get good people on the team, and

TABLE 6.2   Differences between Project and Line Management[a]

	Project management	Line management
Focus	Project; single focus	Organization; multiple foci
Time focus	Short term	Short to long
Attitude	Focus on project	Overall organization performance
Personnel	Close working relationship	Overall management
Budget	Project	Organization
Drivers	Milestones	General objectives

[a] These will be different for each project and organization.

obtaining management support. These are not sequential activities. Project setup is an interactive process in which you work in different areas concurrently. In a sequential process, you would first develop the plan. Then you would get management approval. Finally, you would assemble the project team based on the plan and management approval. It is not that simple. You have to first work on the project and then begin to identify who could work on the team. You should also be communicating with upper management and with line organization managers.

In setting up the plan, project managers should look for examples of past successful plans. Borrow successful ideas. They should also be aware of project failures and learn from these. They should be able to explain what they have learned from these projects (without mentioning the projects by name) and then relate this experience to the project. This demonstrates that they have learned from the past and gives management more confidence in them and the project.

Let's review the goals in setting up the project. On the surface it is to get the project plan developed, approved, and started. But this is a very simple view. We really want to achieve these objectives and also gain concurrence from line management so that they will support the project and provide good people. Many projects get started, but are handicapped from the start because of the bad feelings generated during the planning process. You should assume that you have no power and that you absolutely need the support of key line managers.

You will have to talk to line managers and attempt to attract good people. Through your knowledge of the organization, you should be able to identify who might be good candidates so that you will not be stuck with whomever is available. You should explain the project to the line manager and show the importance of their organization to the project as well as how well organized you are. Do not appeal to the importance that management attaches to the project. This should be self-evident without being stated.

Assuming that this goes well, the next step is to identify the skills that you require. The line manager may ask if you have specific people in their organization in mind. You do not want to tell them how to manage, but you can make some suggestions. If you identify some key staff, indicate that you will only use them for critical work and that other people on their staff might do in general.

Setting up the plan may also involve establishing web pages of lessons learned and issues. The project then demonstrates the suggested pattern of behavior of use.

——————— DAILY LIFE OF A PROJECT MANAGER

Although every day in a project is unique, a brief representative glimpse of a typical day in the life of a project manager might be useful. We assume here that a project is underway and that there are no burning issues to be addressed.

At the start of the day, the project manager should review the status of the project and go over the current issues. What can be done about the issues today? What follow-up is possible? Because any actions may span the day, it is important to identify any actions the previous day or at the start of the current day. Map out what you have to do that day. This will drive the rest of the schedule that day.

Actions on specific issues may mean arranging and conducting meetings with managers, line managers, and members of the project team. You will require time to get organized for these meetings. Make sure that you go into the meetings with a grasp of the situation in your mind. While you should carry some project information for reference, including the project plan, you should make the effort to have it in your head. Go into each meeting with a defined set of actions or things that you want to accomplish. Feel free to state what you hope to accomplish at the start of the meeting. After each meeting, identify specific follow-up steps and make notes.

But these activities may not fill up the day. You should visit at least one person on the project team so that every two weeks you will visit each team member. Sometimes, this can be done at lunch or during breaks, but do not make a habit of using lunchtime to gather information. Most people do not want to talk shop during lunch; they want to eat, run errands, read, and so forth.

Other activities that occur are actual work on tasks, doing project analysis and updating the plan, reviewing key issues, and preparing for project presentations. These should be scheduled in your office at nonpeak times (lunchtime and in the late afternoon).

At the end of the day, you should review what you accomplished that day. Do not wait until the next day as your time will be consumed the next day and you will never have time to get a view of the larger picture. After doing the review, start planning for the next day. What you want to develop is a feeling of accomplishment. Realize that some days there will be frustration. Meetings were canceled; meetings went nowhere; new issues arose; old issues resurfaced, and so forth. That is part of project management.

TAKING OVER A PROJECT

In many cases you do not have the luxury of starting a project from the beginning. You are assigned to take on an ongoing project. We address taking over a failed project later. Here we address taking over a project that is underway and that has its normal share of successes and problems. What do you do first? You will be given the assignment by a manager. Ask the manager what they perceive to be the state of the project and what the issues in the project are. Ask them also what they expect in terms of short-term results.

Meet with the previous manager and take over the files and plan. Have them explain the plan but focus on the issues and outstanding problems as well as the current state of the project. You will be expected to land running in the project. Keep the first meeting short. Read and review the material. Then have a second meeting to go over details. You should probably assume that you will have only two meetings. After all, the old project manager wants to spend time on their new work.

Now you have basic project information. It is time to meet the project team. Go to each team member individually and introduce yourself. Ask the team member to explain what they are working on and what they see as the issues in the project. Ask them what they need or where they require help. This will establish rapport with each person on the team. Do this for people who have even small roles on the project. Consider, if you have time, contacting people who used to work on the project and get their views.

Next, visit the line managers who have supplied people to work on the project. Again, seek out their concerns and issues. Tell them what you intend to do next and that you will get back to them after you have taken over the project so that you can give feedback.

It is now time to have a team meeting. Prior to the meeting you should identify the top ten or so things that you need to address as a manager. Get concurrence from the team on these issues. Do not disrupt any ongoing work. You must instill a feeling of calm.

After the meeting, start to design any changes to the project plan that are needed. Do not implement any changes until they have been thought through. It is more important to determine the true status of the project and start addressing project issues. This shows the team that you are action oriented.

Implement changes one at a time. Don't stretch it out. Otherwise, the team will feel that the project is not stable. Make sure that you present the reasons for the changes with the changes themselves.

CONFLICTS WITH LINE MANAGERS

There can be many different conflicts and disagreements between the project manager and the line managers. Who is to be assigned to a project is one issue. The timing of project availability is a related issue. Fighting over budget and other resources is another source of problems. We do not want to enumerate all of the potential conflicts. Instead, we want to explore what is behind the conflict so that you can understand and deal with the problems.

We have shown the differences between the project and the line management. People are coming from different directions and are acting out of different interests. But it goes beyond this. Some line managers fear that the project manager is attempting to build an empire. This is a reasonable concern if the project is large. The project could be larger than some departments. People may perceive a threat. How do you deal with it? Identify your goals and show by your actions (more than words) that you do not want to build an empire. You can do this by involvement of the line organizations. Give them credit where it is due.

To avoid problems you also need to show how you view the project. You should view the project as a temporary affair. The project is a means to an end. After the end is achieved, the project goes out of business.

TURNING AROUND A FAILING PROJECT

Suppose that you have just been appointed as the project manager and that the project is in trouble. You will be following the steps we have identified earlier. But there are some specific things that you need to do. First, you should gain consensus among the members of the project team about the problems of the project. Even if the problems are obvious, you should get agreement. Then you should start to work with individual team members on the issues in the project.

A second step, after you have identified and developed an approach for each issue, is to devise an overall strategy. Go back to Chapter 4 and work on a strategy and a revised project plan. Take nothing for granted. You usually have only one major chance to turn the project around. If you make the wrong steps now, you could be doomed. Outlining alternatives is very important here. Try to see how much you can improve through project changes without any additional money. Going back to ask for more money is not going to be popular. The project could be killed. You may have to ask for more time. We will address the details of change in Chapter 11 along with addressing specific crises in a project.

MEASURING YOURSELF AS A PROJECT MANAGER

How are you doing? Some project managers measure themselves by the state of the project. This is just a start. Here are some key questions:

- How rapidly are issues being addressed and resolved?
- What is the general mood and atmosphere in the project?

- How cooperative and helpful are line managers toward you and the project?
- To what level of detail do you know what is going on in the project?
- How often do you get around to visiting each member of the project team?

HOW TO BECOME A PROJECT MANAGER AND EXERCISES

Being a project manager is a lot of work. But it is also a means to get into management. How do you become a project manager? You should have some project experience so that you are familiar with the methods and tools of project management. You should measure yourself in terms of the characteristics we have identified for the role of project manager.

It is unlikely that you will be selected to be a project manager for a major project without experience. Look for small projects that are low profile and low key. There are projects that have limited benefits and little risk. You could nominate yourself for one of these projects. What if there are no projects available? This happens often. Look around your department and think of ways you could improve efficiency or streamline the work flow. This should give you an idea for a project.

What do you hope to accomplish with the project? Certainly, you are not trying to make a big impression. What you are trying to do is see how much you like being a project manager and you are also trying to demonstrate your abilities to management.

TWENTY WAYS TO FAIL AS A PROJECT MANAGER

Table 6.3 gives a list of twenty ways you can fail as a project manager. Keep this list in front of you and refer to it often. Note that failure often creeps up on you. It will not suddenly appear. It has been in the making for some time. There are also usually multiple causes for failure. So you have to keep an eye out for symptoms.

TWENTY-FIVE WAYS TO SUCCEED AS A PROJECT MANAGER

Table 6.4 gives a list of twenty-five ways to succeed as a project manager. This is another list to keep in front of you. When things are depressing and not going well, look at this list and see how you are doing things right.

TABLE 6.3   Twenty Ways to Fail as a Project Manager

1. Take a hands-off approach to project administration.
2. Do not get involved in individual tasks.
3. Let issues drift and remain unresolved.
4. Be unwilling to listen to suggestions for change.
5. Be overfocused on specific project management tools.
6. Become obsessed with percentage complete for tasks.
7. Measure milestones by presence and not quality.
8. Devote too much attention to relations with management and not enough to the project team.
9. Be overconcerned with project administration and neglect project management.
10. Attempt to micromanage the project and not delegate.
11. Be formal in relations with project team.
12. Do not stay in communications with line managers.
13. Make too many changes to the schedule.
14. Be willing to rapidly adopt new tools without assessing the consequences.
15. Be status-oriented and not issue-oriented at project meetings.
16. Develop an overly general project plan without detailed tasks.
17. Be tool-focused as opposed to method-oriented with the tools supporting the methods.
18. Fail to regularly communicate in person with all key members of the project team.
19. Leave issues unsolved and allow them to fester and grow.
20. Address issues without analysis.

BEING TYPECAST AS A PROJECT MANAGER

OK, let's suppose that you are a successful project manager. You will likely be asked to manage yet another project. And then another and another. That is good for your ego, but there are practical reasons why the organization will do this to you. First, they have very few people who are good project managers. Second, the may have typecast you in this role.

How do you break out of this situation? Maybe you do not want to escape yet, but if you do, then here are some options. You can go to management and plan for an organized move. You will get more support if this request is made in conjunction with doing another project. The transfer would then occur after the project. Another approach is to establish contact with line managers who respect your work. They can, perhaps, help your mission.

When should you make the transition? Ideally, it would occur at the end of the project. However, if the project is very large, then you might consider doing it at the end of a phase. You should never leave in midstream. It will cast a pall over your work and role. People may suspect something even if there is nothing but success. When you transition out, you should make yourself available to help the next project manager. You should volunteer to be available to answer questions later.

Example   113

TABLE 6.4   Twenty-Five Ways to Succeed as Project Manager

1. Know what is going on in the project in detail.
2. Understand and be symphathetic to project team members.
3. Be able to make decisions.
4. Understand issues and their importance and meaning to the project.
5. Communicate effectively with management.
6. Develop alternative actions.
7. Translate actions into specific changes in the project.
8. Know how to use project management tools and methods effectively.
9. Be able to learn from past projects.
10. Be able to criticize yourself and your performance.
11. Be able to take criticism.
12. Understand trade-offs involving the schedule and budget.
13. Listen to project team members.
14. Understand and act on suggestions for improvement.
15. Be open to new methods.
16. Understand the trade-offs between the project needs and the needs of the organization.
17. Communicate effectively with line managers.
18. Manage your time well.
19. Set up and manage the project file.
20. Be able to generate and use reports from project management software system.
21. Have patience.
22. Be able to take a longer term perspective.
23. Have a sense of humor.
24. Relate current events to project management and the project.
25. Be able to run a meeting.

EXAMPLE

The project leader in our example was someone who had substantial hands-on experience in the manufacturing process and had also worked as a project scheduler and lower level manager. The manager took steps early in the project to establish rapport with line managers for not only the integration and final assembly area, but also other areas. Another step was to take a very low-key, low-profile approach for the project. Anticipated results were minimized so as not to raise expectations. Another step was that the project manager stressed continuity with the past, even though the new project management process represented a revolution over the past. Managers and staff involved in the first project (integration and final assembly) were sought for widespread participation. Initial project tasks included group meetings where area managers and staff could identify all of their problems and gripes in planning sessions. This helped build consensus and will be explored later in more detail.

SUMMARY

We have covered a lot of ground in this chapter. We have tried to show that project management is much more than administrative or standard task planning work. It is exciting and can be enjoyable if you like to solve problems and work with people to accomplish a goal. Project management is not for everyone. But more and more people will be doing it as organizations empower employees and then hold them accountable for getting work done. We have also portrayed the project as separate from the line organizations. Projects can obviously be performed within the line organization. What we have said still applies.

7   The Project Team

A FEW WORDS ABOUT THE WORD *TEAM*

In projects, all of the people associated with a project are working toward the achievement of a common goal. Because of the extent of work, the demanding schedule, and different skills needed, most projects require more than one person. Let's expand on the concept of a team by identifying how they will be working together.

1. They will be working on tasks that involve more than one person.
2. Some people's tasks will require milestones or end products from the tasks of other people on the team.
3. The team will share common methods and tools.
4. The team will have to identify and solve issues together and live with the results (together).
5. Most importantly, the project will sink or swim depending on the final end product. If one person screws up, the team suffers.

It is the interdependence and accountability for the final result that makes it in everyone's self-interest to have the team succeed.

TEAM MEMBERS—DIFFERENT MEMBERS OF THE CAST OF A PLAY

In a play we have several different types of players:

1. The backers of the play—management, producers, and investors
2. Director—the project manager
3. Main players—people who remain in the play during all or almost all acts and most scenes
4. Bit players—people who come into the play for specific scenes and roles and then disappear
5. Cameo players—often, big-name stars whose appearance adds luster to the play; management in doing reviews is one example of a category of cameo player.

Why are we concerned with the cast? First, the traditional concept that all team members are full-time project members is increasingly obsolete with corporate downsizing. There are fewer good people and they have to be spread around on different projects. You also don't want to have nonproductive people on your team and be stuck with them. A second reason is that to attract good people we have to define the timing and a minimum amount of time to allow them to be deployed on other projects. If we ask for too much, we will get nothing or a runner-up person assigned. Third, you want to bring in fresh blood and a new outlook. Otherwise, the project can become inbred and out of touch with reality.

ISSUES REGARDING TEAMS

Now that we know why we need a team and the types of team members, we can identify and start to address key issues regarding teams.

- Where do we start in forming a team?
- How large a team should we have?
- How do we assemble a team?
- How can we attract good team members?
- How do we manage the team?
- How do we bring on new team members?
- How do we remove team members?
- What types of problems arise in teams? How can we deal with these?

GETTING THE TEAM

As we stated previously, there is a natural competition here for team members. That is, the project manager has to attract people from the functional and line organization of the company or agency. These people have current projects and work that will compete with your demands.

Let's assume that you have the project objective and general schedule defined. We will also assume that you have thought about the methods and tools that the project will use. Sit down and consider project risk. Among all the things that have to be done, where is the greatest risk? It is not just in some arcane technical area. People often overlook the following areas:

- Getting a lot of routine design or semitechnical work done quickly. You are after fast and good performers.
- Being able to integrate pieces of a project and test them or evaluate them. Integration is a very difficult skill to acquire and master.
- Good writers. If you have a project that requires substantial documentation, then you want to find good writers.

How do you define what you need? From our discussion in Chapter 4, follow these steps:

1. Lay out a general task plan of 50 tasks, if you have not already done so. Make sure that these tasks cover the project.
2. For each task, identify the skills and knowledge you need. Label each task with the department where you are likely to find the resource and the skills that are critical.
3. Now think about dependencies between the tasks, extent of effort, and the scheduling of the tasks. You are trying to develop a general idea of how much effort you will need and when it will be required for each task.

When you have completed these steps, you should review them with management to get a reality check. You should have identified a core set of players for your play. Off on the side you have also identified the bit player roles.

PROJECT TEAM: THE POLITICAL DIMENSION

You must ensure that the project team contains people from all organizations that are involved or impacted by the project. This might include administrative groups such as internal audit, information systems, and general services (or general administration), as well as line organizations. If the manager indicates that they do not need to be involved, you can accept this, but return periodically, give them an update, and offer them participation again.

TEAM SIZE

From many projects we know that the golden rule is to have a small team. However, too small a team may mean that critical skills are missing or team members overworked. The advantages of a small team (and, hence, the disadvantages of a large team) include the following:

• Coordination is easier with fewer people.
• There is less time spent in informing people of what is going on and in getting status updates.
• People tend to have more flexibility and greater accountability because they have more tasks individually and do more tasks themselves.
• The project manager can be involved in doing some of the work.

We also know from many projects that the larger the project team, the more likely the project is to run into problems and failure. Communications overhead is directly related to team size.

GETTING THE TEAM MEMBERS

You have to think about the personal attributes that you seek. This is a trade-off. Some of the key attributes are listed below.

1. Experience relates to having done similar projects in the past.

2. Knowledge relates to the experience and expertise with the methods and tools. Remember that if you bring in a person who is a tool expert, they will attempt to use the tool on whatever you give them. This is specialization.

3. Business process experience is often critical. You are looking for a person who knows how the eventual user of the project thinks and works.

4. Problem solving ability. If you have a complex, fuzzy project, then you may want people who are good at problem identification and solving.

5. Availability is a double-edged sword. Often, the people who are most available are the ones you do not want. However, if you are given someone who is already overcommitted, you will not get much.

6. Ambition, initiative, and energy—we have found that these qualities are very rare and that they make up for a lot of shortcomings in other areas.

7. Technology. Beware of people who know too much about the technology. They may be technology dabblers—people who like to study it, but who cannot get down and do the work. Can you afford technology appreciation?

8. Communications skills. Everyone wants this, but the people have to work and not just communicate. Some people are great communicators, but do not perform.

After thinking about what you want and what skills you seek, try and find out through the ''grapevine'' who is good, who is available, and who might want to be on your project. In some organizations, you might interview them directly. In larger organizations you have to go to a functional manager and request the people. For example, you might prefer someone who is younger and less experienced, but is more ambitious, knows current tools and methods, and is available.

Getting the Right People

Make an appointment to see the manager. Tell the manager that you wish to discuss your project requirements for staffing. Also, point out that you have made an effort to minimize your requirements and that you will bring project documents with you to prove it. When you enter the manager's office, give them the following items: (1) an overall project plan and (2) a description of the tasks and general schedule that pertain to the person from their organization. In the meeting be precise as to what attributes you are seeking and why these are important. Encourage them to suggest names of people from their organization as candidates. If they ask you if you have any one in mind, you might say, "Well, I would like Joan Stow, but I know how critical her work is to you. How about part of her time and someone else?" If the conversation goes this way, then you may be able to cut a deal then and there. If you do, make sure that you, and not the manager, put it in writing immediately after the meeting as a memorandum of understanding. This will save time later. By the way, the person you are requesting should get a copy of the letter.

If, on the other hand, the meeting is not leading anywhere in terms of closure, you should terminate the meeting with the understanding that you will return in a few days to discuss it again. This will show persistence. You are willing to come back every day to get this resolved.

Our final suggestion here in the end is take what you are offered and make the best of it. How they arrived on the team should not affect their tasks and how they are treated in the project. If you give the impression that they were mandated on the project, the team will pick this up. The result may be a crippled project at the start.

Getting Bit Players—The Casting Call

You may have additional needs that are less than full or even substantial part time. Line up the type of people you need by project phase. Our suggestion from our experience is to staff for two phases ahead. Further out, there is too much uncertainty in terms of effort and schedule. Although the process of getting these people is generally the same, there are some differences.

1. You want specific skills or knowledge so that is the focus of your search.
2. Be clear that the position is not full time and that there is no possibility of staying on the project.
3. Do not raise the expectations as to rewards. Bit players are not rewarded well. It is part of their job and compensation. Like hired guns in the Wild West.

TIME ON THE PROJECT

Traditional project management assumed that people were assigned to the project indefinitely and full time. This is a luxury we can no longer afford. Plan at the start of a project on what criteria you will use to release resources. Examples of criteria are:

- Elapsed time on the project
- Completion of work
- Turnover of work
- Requirements for their time from higher priority projects

YOUR FIRST MEETING WITH INDIVIDUAL TEAM MEMBERS

A team member has been identified through some means. Let's face reality. The person is likely to be nervous. They will have some real concerns and questions. Here are some we have encountered:

- What is expected of them?
- What is the project environment?
- What will happen to them after the project is over?
- Why were they chosen?
- What will happen to their old work?
- Will they learn new skills?
- How will the project help their career?

You have to address all of these concerns. Even if they do not voice these questions.

Let's go through the meeting. First, it should be in their office. You want to get the lay of the land. What is out on their desk? What are they working on? You sit down. Start right in. Give some reading materials to them to review after you leave. This should consist of the same materials that you provided their manager. You might also add any details on their tasks and information on tools and methods. Now it is time for your presentation. Start with the project overview. Then zoom in on their tasks. Explain why their work will be important to the project. Go into why they were chosen. Explain what skills they have that you value. Discuss the methods and tools. This detailed discussion will help ease any concerns. With that anxiety relieved, move back to the project as a whole. Go over the schedule, problems, and pressures. They will begin to feel that they are a part of the project.

To summarize, here is a list of items you might consider for your meeting:

1. Objectives of the project
2. Background and history
3. Structure of the project
4. Project milestones
5. Their role
6. Their tasks
7. Their milestones

8. Organizations involved in the project
9. Challenges and issues
10. Methods in use
11. Tools in use
12. Why they were selected
13. Expectations of work
14. Their schedule
15. Evaluation of their work
16. Process for handling issues
17. Status and monitoring process
18. What the project can contribute to their career
19. Criteria for release from the project

Now that they understand the project in general and their tasks in particular, you are ready to move on to the controls and reporting of the project. Discuss how the project is being managed. Discuss the other team members and their roles. Cover how issues will be discussed and addressed. You want to empower the person and encourage them to be accountable for their detailed tasks. Get them to start planning their work.

Set up a time for the next individual meeting with them. At the second meeting you will go over any questions. You will also develop the schedule for their tasks with them. Schedule this meeting only a few days after the first meeting.

YOUR FIRST TEAM MEETING

You have the team identified and on board, but they have not assembled as a team. The first meeting is a big event in the project, but you should not treat it that way. It may give false signs or raise expectations. In this first meeting, you are trying to accomplish several goals:

- Have them meet each other in the context of the project.
- Repeat briefly the objectives and scope of the project.
- Describe the general schedule.
- Go over some initial issues.
- Discuss the methods and tools.
- Simulate an average meeting to show them what future meetings will be like.

Keep the meeting short—less than one hour. Select a location that will be typical for the project later. Do not have coffee, food, or other distractions. Begin with an introduction in which you indicate what each person's role is. Let them describe their background in their own words. With an agenda for the meeting, present each person with some handouts related to the project, tools, methods, and issues. Get right into the topics. Encourage questions along the way. Have each identify themselves and describe what their role in the project is—in their own words. You now make the next meeting and later ones easier. You need to get people comfortable with the process and with each other.

KNOWLEDGE AND USE OF NETWORK BASED TOOLS

Team members must be familiar and at ease with the network based software tools that will be used. If necessary, provide not only training and demonstrations, but also guidelines on how they can be used most effectively.

MANAGING THE TEAM

In Chapter 8 we will cover monitoring of the project. Here we will state that our approach is to obtain status information informally from each team member and to use meetings to address issues. As much as possible, you handle problems outside of the meeting, and you simply tell people the results of the resolution and actions. With status out of the way temporarily, we need to address communications within the project team.

Some specific issues are how to convey information, how to address problems and opportunities raised by the team, how to deal with conflict within the team and between the project and functional departments. We will start with getting information out. If you have electronic mail or groupware, start a group under the mail system for the project team. This will be your main public way to get the word out to the team. Alternatively, you can use voice mail and the memo approach. Having set up this system, do not deviate from it.

ADDRESSING PROBLEMS WITHIN THE TEAM

Table 7.1 gives a list of potential problems and opportunities that can arise. Obviously, the list could be infinite so that those on the list are only to be taken as examples and as being representative. Issues can arise directly by someone in or outside the team pointing out a problem. They can be unstated, but we all know that they are there. When you have a problem or issue presented, you should first have it discussed. Do not attempt to deal with it. You are trying to understand it. You are trying to decode the following:

- Symptoms of the problem
- Aspects of the problem itself
- Relation of the problem to the organization

TABLE 7.1   Examples of Problems and Opportunities within the Project Team

Early alert and identification of issues
Identification of new methods for use in the project
Identification of new tools for use in the project
Conflicts between line management and the project over project assignments
Conflict over specific methods or tools used
Project restructuring opportunities for greater parallel effort
Competition over authority on resolving issues
Conflict on specific issue solutions
Resurfacing of past issues
Inability to work together on specific tasks
Priority conflict among specific tasks
Team member reassignment from project
Physical illness or vacation
Lack of interest in work and project
Sense that project is not going anywhere
Disagreement over project objective, scope, and strategy
Conflict over structure of the project

- Parts of the project impacted by the problem
- Impacts if the problem is ignored

Do not attempt to resolve the issue there. Collect the information and then do your own analysis along the lines we will discuss later in the book. In doing the analysis you should solicit ideas from the team. When you have a resolution or need to discuss it in a group, then bring it to a meeting of the team.

Here are several tips. First, run the solution by the team members who will be impacted most. Get their support and understanding. The solution will not take them by surprise when you present it at the meeting. Our second tip is to have them present the solution to the team.

Conflict within the Team and Externally

Conflict can be open or an undercurrent in the project team. Our first suggestion is that you learn how to recognize and deal with it. Recognizing it does not mean waiting for open conflict—look for tension and watch body language. Dealing with it does not mean resolution. It may be that the conflict is too deep-seated and will long outlive the project. You are not out to solve organizational conflict; you are there to complete a project.

What is the source of conflict? Much has been written on this in an academic sense. From what we have experienced in real life, some of the major sources are as follows:

1. Money. People are always fighting over money; this is eternal. This is also one of the easier conflicts to address because it tends to be more rational.

2. Priorities. This is conflict over where people spend their time. It is often a symptom of a deeper issue that questions the work of one project in relation to another. In the battle between the functional and project managers, priority is one of key battlegrounds.

3. Schedules. Conflict occurs here in almost all projects. It is the struggle over how long the work will take to be done.

4. Control. This can be control over the project and the role of the project manager. At the heart of this is the authority of the project manager.

5. Procedures and methods. This is an argument over how the work is to be done. This can get very heated and it can question the ability and skills of the team member—just as control issues question the authority of the project manager.

6. Personalities. This is the conflict of personalities in the project team. We have saved the worst for last. Of all of the problems and conflicts, this is the most emotional and has the least to do with the project. If the people involved in the conflict are important, then you have to live with it.

How do you deal with conflict? You need to understand it. Several alternatives are resolution, no action, deferral, and transformation or change of the conflict basis. Deferral and inaction buy time and allow for the situation to sort itself out (or get worse).

One way of dealing with conflict is to impose a solution. In the case of personalities and other conflicts that will exist and outlive the project, we favor an approach where the different points of view are recognized and put on the table. This will serve to minimize the damage and fallout from conflict.

The goal in conflict resolution is not to win. You want the project to win. Therefore, your goal is to minimize the impact of the conflict on the project. You can do this not only by acknowledging the conflict, but also by reorganizing the project so that conflict is kept to a low profile.

When we speak of conflict, a number of different images come to mind. Some think of war; some think of argument. This is normal. What we are seeing is the modifier to conflict—intensity. In order to minimize the impact on the project, we can work on lowering the intensity of the conflict.

But is all conflict to be avoided? Not at all. As a project manager you will have to confront and win conflicts over resources. You don't have to win every conflict; the project needs to win. Our guidelines come from Caesar's written records:

- Choose the arena and issue of conflict carefully. If you choose the arena, you are more likely to win.

- Fight a conflict with information and rationality on your side. If you resort to emotion, you will lose.
- Choose issues that are of true importance to the project. If they are important to you and not to the project, be careful. You risk losing a lot. Also, there is a fundamental problem—namely, that your position is not congruent or aligned to the project. You are out of step. Get in line.

Conflict in projects in the 1990s and beyond is natural because the projects tend to have substantial impact on the organization. Expect it and proactively deal with it. Even if this means acknowledging it and then putting it aside.

GETTING MORE OUT OF A TEAM MEMBER

We have indicated that you want to give positive feedback and attention to people who are doing well. What about the flip side? How do you detect and handle problems in performance and motivation? Although this is not a psychology book, we can offer some observations from the past.

1. The first signs of problems will not be in missed schedules and poor work. It will be seen in the member's eyes, facial expressions, voice, and body language. That is why you have to visit them in their offices—to see what is going on.

2. Assuming that you sense a problem, ask about their overall work and tasks—beyond the project. Try to get them to open up. If they do, try to help. Do not run and tell someone what they say. They are taking you into their confidence. The walls have ears. Betray that trust at your peril.

3. Carefully review their tasks to see what you and others can do. You do not want to relieve them of all tasks. But you do want to give them some relief. Our experience is that this act of kindness will be repaid many times later.

What if they are stuck on a technical problem or disagree on a technical approach? Suppose they do not want to admit their lack of knowledge. This is very typical of junior staff on technical projects. They will spin their wheels because they are intimidated by the senior staff. How do you deal with this? Address it before it happens. Early in the project, indicate that it may happen. Then assign a junior person to a senior person for assistance. This goes back to something that most of industry has lost—apprenticeship.

BRINGING IN SOMEONE NEW IN THE MIDDLE OF THE PROJECT

As we have seen in today's projects, there is more turmoil and less stability. This translates into more turnover in the project team. How do you deal with turnover in general? At the start of the project, indicate that this is what happens and that you hope that there will not be too much. But it will happen.

How do you bring someone new into the project? There are two cases—planned and unplanned. Let's look at the similarities in the situations. You need to assemble an up-to-date set of materials and give it to them. You need to have the meeting along the lines suggested earlier.

After they are in tune with your explanation, introduce them to the other project team members one at a time. Start with the people that they will be working with closely on tasks. When you introduce them, include the following:

- What skills they have and why they have been chosen
- What they will be working on in terms of tasks
- What the existing team member does and why it is important
- How the two people will be expected to work together

You are trying to build rapport and establish mutual respect between the people. You should assume that the new person is wondering why they are there. The current staff may be thinking that the new person represents a threat to their position. You are trying to head off problems and instill calm. As a project manager, you will see that you spend a lot of time mixing oil and water.

Some additional comments are useful in the case of the person who is being brought on board unexpectedly. They may be coming on board for a number of reasons:

- Their skills are needed.
- They are replacing someone who has left the project.
- They are excess and have been placed in your project.

The last one sounds pretty bad, doesn't it? It is not. The person probably doesn't want to be there. You should sit down with them and go over the material we mentioned. You should also acknowledge that this project may not be their first choice. Get it out in the open. Deal with it. Tell them that you insist on performance and being a team player, but that you are sensitive to their feelings. Have them identify for you how they can best be of service. This will get their commitment. After all, you are trying to make the best of a situation on their behalf and on behalf of the project.

You should pave the way for this new person by telling the team members individually about the person and their skills. It goes without saying that you should not mention if they are being placed in the project. Everyone knows anyhow. The rumor mill is faster than electronic mail.

GETTING RID OF A TEAM MEMBER

There are times when after counseling and working with a team member you find that they can no longer function as part of the team. You must embark on two parallel efforts.

- Undertake Damage Control. Move most of their tasks and certainly all critical tasks to others. Or, take them on yourself. In our last chapter, we are thinking of you as a hands-on manager and doer. Not an administrator. If you want to administrate, get out of projects and go into functional, line management.
- Go to Management and Orchestrate a Transition Plan. Do not put pressure on management. They will not appreciate another problem. That is why we said in the last chapter that you need to solve as many problems yourself. Regard appeals to management as the Lone Ranger regarded silver bullets. Use them very sparingly. Assume that the transition will take some time—maybe several months.

What do you do if the person has to remain in the project? Do not regard the situation as lost. View it as a challenge. Try to assign them less important tasks. All projects have unlimited tasks. Use your imagination and creativity.

What do you tell the rest of the team? They probably already know. But they are watching to see how you, as the leader, will handle it. You need to demonstrate style and compassion. They may be wondering if this will happen to them. You should tell them individually. Do not address it in a group unless someone asks. If you do it off-line, individually, then it should not arise. If it does come up, admit that it is being worked out and that you will keep them updated.

───────── DEALING WITH DISASTER

Let's suppose that a milestone is not reached. Or that the work does not pass acceptance. Assemble the team and go over why it happened, what can be done to recover, and how to prevent it. Do not place blame. It will make matters worse. However, you should shoulder the blame with management and in front of the team. You are the manager. You are accountable. How you respond to crises counts. Anyone can deal with everyday events.

You should also analyze the disaster from different points of view. Look at it from management, the project, the team, and other perspectives. What were the symptoms that you missed? Why didn't you pay attention? Did it happen because you trusted and believed what people said? We all have to trust other people because we cannot do everything ourselves. Do not stop trusting. But take precautions. Look for proof of statements about milestones and progress. This is especially true about progress on tasks. That is another reason why you need to be personally involved in tasks. By getting your hands dirty in the project, you will know more of what is going on in the project.

RESTRUCTURING A TEAM

With the changes in the project and the team, it is probable that the project will have to be restructured. Do not wait until events dictate restructuring. This is a reactive mode. Act quickly enough. England made many mistakes in World War I with commanders in the field. They failed to replace them quickly. Hundreds of thousands of lives were lost as a result.

Tell the team at the start that you will have to restructure the project from time to time. By making fewer major changes on an organized basis, there will be less disruption than reaction or having many, smaller changes.

How should you proceed with restructuring? Begin with the plan and revise it first. Evaluate the methods and tools. Look for a bundle of related changes. As for the team, you should restructure in an orderly way that does not upset current work. Reassignment of tasks and assignment of new tasks should be phased in and not forced immediately.

How do you present the new structure? First, present it to management by stating your objective and strategy. Go into the impact on the budget and schedule, if there is any impact. Most of the time, there will not be a major impact. So, it is likely to be approved. They want you accountable and you are showing how you are dealing with the situation.

Second, you should go through the same presentation with more detail in a team meeting. Be ready to answer any and all questions. It is likely that the team members will bring up issues and questions that you have not considered Thus, keep the fine details fluid. Accept their input and change the plan within the strategy and objective. In that way, they will accept the changes more readily. When you have the next meeting, go into how you have taken their comments into account.

HAVING FUN

We have given you a number of ideas for handling situations and problems. Now it is important to have some fun, for their sake, for the project, and for you. Here are some ideas:

1. Circulate copies of articles that describe project disasters and problems. Your problems should pale by comparison.
2. Look for cartoons and jokes about projects. Encourage the team to find some. Start circulating them.
3. Develop a body of historical examples from large projects. Refer to these as we have done in this book. These examples will give perspective.
4. In the project, indicate that at least in this age, the project team and manager will not be shot. Dark humor, but it can break the ice sometimes.

TAKING SUGGESTIONS FROM THE TEAM

What do you do with suggestions voiced by the team? First, you should understand and listen to the suggestion. Why is it being made? What is behind it? Why is it being suggested now? A rule that we have lived by is that very little happens in a project randomly. There is a reason that the suggestion is being made at this time.

A second step is to solicit any additional, related suggestions. Having gotten these out on the table, do not respond with answers when they are made. Go away and think about them. What can you do to implement the suggestion? Take a positive view and assume that you want to take action. Do not dismiss it. If you do, there may be no more suggestions. That means trouble and a lack of communications.

REWARDING THE TEAM

We discourage setting up bonuses and rewards. In projects where this was done, the results were counterproductive. People who did not receive a reward feel slighted. People who are rewarded want more and begin to expect it. Good work should be recognized. You should feel free to send a note to the person's manager on how well they did their job. We are assuming that they made an outstanding contribution. If you acknowledge mediocrity, you will end up with a mediocre project.

AN EXAMPLE OF TIME

The role of the clock in the advance of civilization has not been widely known. Development of the clock and time pieces made possible precision down to the hour for work. The word "punctual" came into use. As people wanted personal clocks, clock makers had to design and make smaller components. The manufacturing process was more complex. With this experience, clock makers were called on to make other types of instruments. Ferdinand Berthold in the middle 1700s faced such a problem. Clock making was advanced and the work was more specialized due to miniaturization and the different skills required. Berthold created a project team. What was once a loosely organized project team now became a team of specialists with division of labor. Individual team members were gilders, engravers, finishers, crafters of springs, pendulum makers, polishers, and so on. This example shows how important the organization of the team is in a project.

EXERCISES

Look at a project around you. How are team members treated? Is there any humor in the project? Ask yourself these and other questions that reflect what we have discussed. Even if no problems are evident, ask yourself how the project team could do their job better.

EXAMPLE

A key to success in the project was to realize that with the ultimate large scope of the project and even the size of the initial project, the core project team had to be kept small. Having a large team over an extended period would have created problems in transitions. Second, due to the political nature of the project in terms of reengineering, the project team would be under the microscope. The fewer people under scrutiny meant that there was less miscommunication.

The actual team consisted of two key employees—a management assistant project leader and a technical person for the software and network, a former employee who was familiar with the old scheduling methods, and a consultant who provided managerial expertise and outside knowledge. Two organizations were asked to provide part-time team members for political reasons. These people were placed in less sensitive support and administrative roles. In addition, there were many part-time players from the line organizations for each phase.

Weekly sessions were held with the core team to determine how things were going and to deal with the political issues. These meetings never overlapped with the technical or operational work. Regular communications between core team members was a key to success.

When one of the team members who was hired for political reasons turned hostile and was ineffective, the project leader moved to isolate damage, but also permitted the person to diminish themselves in the eyes of others by their criticism. The person eventually wanted to be off the project. Their wish was granted and another, more cooperative person was placed on the team from the same organization.

SUMMARY

Managing a project team is very difficult and challenging. These people do not report to you outside of the project. You have little or no power to pay them more money or give them a promotion. You have to work with the functional, line managers to see that they were rewarded. We stress organization, fairness, compassion, understanding, and involvement. And a sense of humor.

8 Intranets, Internet, and the Web

A major theme of this book has been the use of modern technology to support projects and project management. In various chapters hints are given on how to use specific features and capabilities of electronic mail, the Web, Internet, and other technology. Here, we explore technology trends and their impact on projects and project management.

There has been a fundamental change in the direction of technology in the past five years. Previously, technology was aimed at the individual working with a large, medium, or personal computer. Helping individuals to be more productive in their work was the theme. This changed with the onslaught of the Internet and Web. Attention has shifted to networks and all of the implications of this technology.

Network technology has the following impacts:

- Information and hardware resources are intended to be shared.
- The benefits of network technology target the productivity of the group involved in a project or business process, rather than of individuals.

- Organizations prefer networked systems and computers since they can be supported at lower cost.

This technology has caught on faster than the automobile. Witness the growth of the Internet, Web pages, and client-server computing.

Businesses have picked up on the above impacts and employed client-server and intranet systems to support reengineering business processes and rightsizing organizations. The keys to success here appear to be:

- Networks can link all of the people involved in a business process so that work can be managed, tracked, and measured better.
- Business processes can be more integrated and, hence, more effective.
- Companies can expand to global enterprises with network technology.
- Organizations can attempt much larger projects that involve more locations, people, facilities, and equipment since the network can extend project management and control.

MODERN MANAGEMENT IN A NETWORKED SETTING

Pick up most project management books and you will find the standard standalone project management approach. That is, you are put in the position of the project manager. You have specific tools and you do work yourself. The methods and tools are standalone. You develop the schedule; you present the schedule; you update the schedule. This approach leads to major issues and problems, including:

- Each project manager works alone and develops approaches that are not compatible with those developed by other managers. People who work on different projects are confused. Individual style and knowledge of the project leader dominate.
- The project plans and schedules generated by the project leaders cannot be put together into an overall schedule since there is no compatibility.
- Each project has to be directed and managed standalone. This makes it hard for resources to be shared.

The end results are that projects are more expensive than they should be, take longer than necessary, and fail to learn lessons from the past. All of these factors reduce the benefits of project management.

This book takes a different direction. In a network setting, project management is much more of a shared, joint effort among the project team. People share information on the project through the network. This fosters a more open environment with more give and take outside of the technology. The following table provides a comparison of standalone and networked project management.

Aspect of project management	Traditional	Modern
Project information	Individual	Shared
Project management software	Standalone	Network-based
Other software	Collection of single user tools	Network-based integrated or interfacing software
Handling of issues	Discrete meetings	Continuous effort to investigate and resolve issues
Project manager role	Substantial administrative burden	Managerial and coordinating role
Project communications	Personal meetings	Electronic communications for more mundane work
Resource management approach	Resources dedicated to projects	Shared resources
Management	Control of individual projects	Control of multiple projects
Project management	Batch oriented through sequential, periodic reviews	On-line oriented closer to real time

The benefits to management of modern project management are compelling. Resources can be released from one project and assigned to another more easily. Management has more power over the project because they have access to more information. Project team members also have more information and can even participate in updating the schedule and defining details for their own task areas.

Should we be surprised at this turn of events? Absolutely not. In the early days of the Roman Empire commanders were left to their own resources in their expeditions. It became very clear to Rome that this provided no flexibility for defense or for development and taxation of the people. Roads, horseback message systems, and standards all helped link parts of the territory together in a network. It is no secret that the Roman Empire succeeded for hundreds of years because of its network. The same was true with the telegraph, which allowed central control of remote activities and revolutionized warfare, commerce, and government.

What is the impact of the modern project management approach on the methods we have discussed in the book? The answer is that networking and modern project management extend the methods. We still must have the same skills, knowledge, and abilities. On top of that, we must be familiar with and adept at taking advantage of new technology to support our project. After all, we often embrace new technology within the work of the project. We should not expect to keep project management itself locked in the past.

AREAS OF TECHNOLOGY

Let's examine specific areas of technology that apply to project management and discuss their application, impact, and potential for abuse. This last area is important because technology can be abused and rip up a project as well as support it. Table 8.1 gives some of the major technologies applicable to project management. Some comments are as follows:

- Videoconferencing and Internet telephone are technologies that work with specialized software and hardware through the Internet.
- Groupware and network database management systems can address the same applications. However, there are some significant differences between them. Database management systems offer the structure of a relational database. Groupware is less structured, but offers more interactive use such as building a document or addressing an issue as a team.
- An intranet is a company internal Internet in that it uses similar hardware and the same software. An intranet provides security since it is not open to all, as is the Internet.
- The Web here is the World Wide Web and the ability to build Web pages and applications based on the Web in Java or other languages.

Our basic point is that you should employ as many of these technologies as is feasible and effective. Here are some guidelines:

- The benefits of network technologies lie in the collection of the technologies as opposed to a specific technology. For example, having a database on issues link to the project management system and vice versa is much more powerful than one of the tools or separate tools.
- How individual and collective tools are used depends on the project, the team, and the style of the project manager.
- A pattern of use must be started soon after the beginning of a project to be effective. It is too difficult to change behavior in the middle of a project on top of the work in the project.
- Guidelines and examples of use should be provided as models. Otherwise, people will improvise and not derive the full benefit of the tools.

What are some trends in the technology that can affect this list? Hardware is improving at a steady pace. This has several impacts. First, it makes the basic technology more affordable to more people. Second, it allows organizations to upgrade their systems. Third, more complex software can be supported with high performance.

In the network area, there have been several trends. First, there have been advances in networking hardware. Second, deregulation of communications has made more services available at a lower cost. Third, networks have more capabilities and functions than ever before.

Perhaps the biggest change has been in the software arena. With a single dominant vendor for much of the market there are greater standardization and greater ability to interface various software. Software functions have expanded. Basic functions are easier to learn and use. Many applications have "wizards" or help systems to show you how to do a particular task.

TABLE 8.1 Network Technologies and Their Impacts

Technology	Application	Benefits	Potential abuse
Voice mail	Collecting project status	Reduces telephone tag	Extensive, long messages
	Disseminating information		People not trained in leaving messages
Electronic mail	Collecting and forwarding information	Obtain files as attachments	Unstructured
	Disseminating information	Collect data for reports easily	Depends on writing skills
Project management software	Project tracking	Team participates in updates	Manager runs project as traditional project in spite of network
	Joint project updating	Team can define detailed tasks	
		Schedule represents truth	
Database management systems	Issues database	Shared information	Database not used
	Action item database	Structured information	Database not maintained
	Lessons learned database		
Groupware	Issues database	Shared information	Software not used
	Action item database	Joint work on documents	Data not updated
	Lessons learned database		
	Project documents		
Calendaring/scheduling software	Scheduling of meetings related to projects	Easier coordination	May not be used by all
			Not kept current
Internet videoconferencing	Issue discussion	Better than e-mail and voice mail	Limited quality of technology
	Technical project discussion	Less chance of misunderstandings	Expense and learning curve for technology
Internet telephone	Standard voice communications	Cheaper than long distance	Quality and compatibility issues
Web/Internet	Web pages and applications	Low cost communications	Requires remote network access
		Flexible	Limited security
Intranet	Provide internal web	Benefits of the Internet	Setup costs
		Secure because it is internal	

Examples are setting up a schedule in project management software and designing a web page. The software products have become network focused and capable as never before.

Overall, what can we expect beyond additional features? A major feature will be the capability to construct a software application that spans several tools. We can do this today, but it takes substantial customized programming. Another trend will be improved performance and capabilities. This will impact videoconferencing and other similar technologies.

THE LEARNING CURVE

As with anything new, there is a learning curve. Here the curve is more complex because there are multiple different tools that interface and interact with each other. With any tool, you begin by learning the basics and gaining experience. Then you progress and your proficiency improves. Later, your mind returns to your normal work and you cannot spend more time learning additional features. You have just reached your first plateau. If you don't require additional features, you can sit there forever. While you won't learn even 30% of the features, you do have enough to get the work done. If you do need more, you learn more of the software and you advance to a higher plateau.

Let's take this discussion and expand it to multiple software tools. In project management we have a very narrow focus. In word processing you learn how to generate documents. For project management, your interests are more narrow. You want to use the software to support you in project tracking, project updating, project reporting, issue management, and other similar tasks. Doing this can require almost all of the software listed in Table 8.1. Here is an example:

- Electronic and voice mail—messages to request schedule updating
- Project management software—team members review and update their tasks
- Review of schedule reveals problems
- Issues database or groupware—review the status of issues that pertain to the specific tasks that have problems
- Action item database—determine what is being done
- Videoconferencing—joint meeting to discuss specific issues

The best approach is to define the steps you would follow in a business setting without the software. This is the logical flow of the work. After you have done this, you can then insert the procedures for each piece of software required. This gives you an overall process. If you build the procedures for the software first, you will have gaps in the process as you attempt to move between the software tools.

MANAGING THE TECHNOLOGY

Before you decide to implement all of this technology for just a project team, let's consider what support requirements there are. Figure 8.1 presents a list of job functions, guidelines, and policies required for the technologies. As you can see, you really do require this support for your project. You would not want to embark on using the technology without support. Why? The project leader and team would end up spending more time on the software tools than on doing work in the project.

There is no way that even a large project team can have its own network management and support. Must projects wait for the technology to be available to everyone before they get a chance to use it? No. A project team can serve as a pilot project for using new technology.

The important things to keep in mind are:

- There must be a critical mass of project teams and schedules on the network in order to derive the benefits of information sharing, etc.
- The use of the network tools will require project management support in addition to systems and network support.
- The approach for moving the teams and establishing the new process must be carefully thought through.

COSTS AND BENEFITS OF NETWORKING TECHNOLOGIES

Unfortunately, the costs of implementing the network technologies are about the same whether you do it right or wrong. Thus, you could implement the tools and receive no benefits or all of the benefits. Some of the major cost areas are:

- Equipment and software components
  Cabling
  Network hardware
  Workstation hardware
  System software (operating systems, utilities, etc.)
  Software tools (e-mail, videoconferencing, etc.)
  Database management systems and groupware
- Staff support
  Network installation and testing
  Software development
  Network support
  Training and procedures support

While many costs are one-time, there are substantial recurring costs in staffing, upgrades, and troubleshooting. The cost per workstation can ex-

Roles:

- Network operations
  Daily operation of the network
  Troubleshooting of problems
  Installing and testing new software versions
  Upgrading hardware components
  Performing network backups
  Restoring files and databases from backups
  Maintaining network operations procedures
- Network management
  Performance analysis
  Performance monitoring
  Capacity planning
  Configuration management
- Security administration
  Password control
  User access control
  Security testing
  Security procedures
  Firewall procedures (securing internal networks from external access and links)
- Internet coordinator
  Protection of internal network
  Web page procedures
  Support for web page and Internet software
- User network support
  User training
  User help desk
  Tracking of problems
  User procedures and documentation
- Developer for groupware or database management system
  Requirements
  Database design
  Forms design
  Software development
  Software testing
  Software documentation
- Project scheduling coordinator
  Training of staff in project management tools
  Review of schedules for conformance to templates
  Development of new templates
  Analysis of multiple projects

Guidelines:

- Using the network
- Accessing the Internet
- Developing Web pages
- Maintaining Web pages
- Project management templates
- Security procedures
- Operations procedures
- Architecture standards
- Network user procedures

FIGURE 8.1 Functions and guidelines required for technology support.

ceed $8,000 per year when all costs are spread over the workstations. This depends on the extent of the network and the number of workstations. This means that implementing the network tools for one or two project teams is not feasible.

Getting benefits from the technologies depends on the extent to which the network will be employed for the following:

- Sharing and access to project plan files on the network
- Sharing of lessons learned and technical information within the project
- Ability to draw information across multiple projects in order to resolve resource conflicts and make management decisions
- Establishment of issues and action item databases on the network and linkage to the project plans

If we are able to implement these, then we would realize savings and benefits such as:

- Improved productivity of project work
- Reduced time in resolving issues
- Fewer outstanding issues
- More certain schedules
- Issues resolved at lower organization levels
- Common schedule in place eliminating people maintaining redundant schedules
- Shorter meeting time and fewer meetings
- Improved quality of work

SOME POTENTIAL PROBLEMS

Over the years we have seen how a number of well-intentioned organizations have failed in their efforts to achieve benefits from networking for project management. We constructed the list below along with some hints on what to do to avoid the problem.

- Problem: Staff do not use the network tools after being trained and directed to use them.

  This can occur several ways. First, the tools could have been implemented in the middle of the project. The project team does not have the time to learn and become proficient with the new tools. A second reason is that the team perceives little benefit and only extra work from using the tools. A third reason is that management proclaims the use of the tools, but fails to follow up with enforcement.

- Problem: Staff use the network tools, but benefits appear to be small.

  This could arise from a measurement problem. If the old process was not measured, then when the new process and tools are in place, no one knows what the benefits are. Another possibility is that people were not trained in how to use the process and tools together.

- Problem: Network tools are in place, but staff feel that the interface and use is too awkward. Usage declines.

In this case there appeared to be no testing of the interface and procedures from a staff point of view as opposed to a systems viewpoint. People become frustrated by the interface and stop using the tools. Staff may also decide to use only a small percentage of the functions.

- Problem: Network tools are in use, but the same project management process is employed.

This occurs when the process was not addressed. People thought that by osmosis or a seventh sense staff would see what new procedures and policies would be necessary.

How could these problems be prevented? The key is to have a strategy in place first that defines the new project management process that employs the tools. The process and the tools then become mutually supportive.

DEVELOPING A PROJECT MANAGEMENT NETWORK STRATEGY

What are the ingredients of a strategy? Here is a list.

- Which projects will employ standardized project templates
- How projects will be set up on the network
- Roles and responsibilities of team members and project leaders on accessing and updating network based information
- Responsibilities and assignment of staff who will monitor and support the new process
- How management will direct and control multiple projects and resolve resource conflicts
- What projects will be allowed to retain the new process
- Timing of the migration to the new strategy

We can now formulate several alternative strategies. One is to adopt all of the potential technologies we mentioned, but leave project management to the project leaders exclusively. This gets all of the costs and little of the benefit. At the other extreme is to implement the strategy without the technology. This is a sure recipe for failure since manual efforts would have to replace the automation. Another approach is to employ the less exotic technology (not using videoconferencing, for example) and move some responsibilities for updating to project team members, resource conflicts to groups of project managers, and multiple project issues to management. This is a middle of the road course. You can prepare a table where the rows are the elements of the new process and the columns are the technologies. A partial example is shown in Table 8.2. In the table, you would place an X if the process requires the technology.

This makes common sense. Why would there be resistance? Project managers may not want to share information since it might mean, in their minds, that they lose power. Information is power. They will lose control. Managers at higher levels might resist because they don't want to become involved in decisions within projects. Staff might resist because they see updating a common schedule as another burden. Trust us, assume that you will encounter all of these and head them off with the benefits.

NINE STEPS IN MIGRATING TO A NETWORK BASED ENVIRONMENT

A key to successful migration is to realize that migration is more than software, hardware, and network. It is changing the culture of project

TABLE 8.2   Project Management Strategy Elements versus Technologies

Project management process element	Project management software	Electronic mail	Groupware/database management system	Internet	World Wide Web
Shared information among team	X	X			
Shared information outside the team	X	X		X	X
Use of standard project templates	X				
Issues database			X		X
Action item database			X		X
Issues database linked to project mgmt software	X		X		X
Action item database linked to project mgmt software	X		X		X
Rollup of multiple projects	X	X			
Joint resolution of resource conflicts	X	X	X		X

management and behavior of the managers and staff. These changes are much more challenging that just plugging stuff into cabling. Here are some suggested steps for migration.

- Step 1: Measure the current project management process in terms of effectiveness of resource use, availability of information, and project controls.

  How do we measure the current process? Here are some things to look for:
  - Age of the oldest major outstanding issues;
  - The degree to which resources can be released from one project and assigned to another;
  - Whether there is any effort to collect and employ lessons learned;
  - Extent to which staff and managers maintain their own separate project schedules;
  - Whether there is the ability to aggregate multiple project plans;
  - Whether projects are managed individually or collectively;
  - Extent to which management has to be involved in detailed project decisions.

- Step 2: Define the new project management process taking into account the network tools.

  Using our discussion of strategy above, we formulate several alternatives and obtain management involvement and approval.

- Step 3: Pilot test the new process and simulate the process without the network tools.

  How can you do this without the tools? Simulate the process with people sitting in a room. Label papers as schedules, action items, and issues. Use bins as the network. Identify someone who will act as a scribe and write down the procedures and results. Conduct a second session where you simulate exceptions and problems.

- Step 4: Refine the process based on the pilot test.

  After simulation and the pilot test, you will uncover issues related to project templates, the cut-off or barrier as to when things move from the network to standalone and vice versa.

- Step 5: Install the network tools and train the staff that were involved in the pilot test of step 2.

  The training in the software should only include the basics. Training in-depth in a specific tool is bad because of the number of tools and because it detracts from the process.

- Step 6: Conduct a second pilot test of the process with the network tools.

  With the procedures from the pilot and the network tools in place, you are ready for the second pilot test. Gather data on an existing project and set up the information on the network. Don't attempt to be complete. This is only a test. The information should be from a real project to be realistic. Also, you will be using this information as the basis for your training of the staff in general later.

- Step 7: Develop procedures, policies, and guidelines for the process and network tools.

  This is a very important step in the migration. It is here that you extract lessons learned from step 6 and define procedures of use. You will also define project templates, set up issues and action item databases, and other activities. Policies are necessary to determine under what conditions the tools will be employed; for example, how severe an issue must be in order to enter it into a database.

- Step 8: Assess which project teams will be moved over to the new process.

  You might not want to convert all teams to the new process. We suggest that you group projects based on sharing common issues and resources. Standalone projects that are about to be completed would be the first candidates for exclusion. New projects would be included. In general, you want to ensure that each team member is working with only one process.

- Step 9: Train and monitor the use of the new process and tools.

  This step includes the measurement of the new process and tools and a comparison with the results of step 1. In addition, you want to use this opportunity to identify any lingering resistance to the new process.

COMMONLY ENCOUNTERED PROBLEMS

Some common problems that people have encountered with network technology are listed and discussed below. None of these is a show stopper. However, each can impact project performance. What should you do about this? Develop a backup plan which does not employ the network technology. Test this plan now and then. This will show that you have an alternative as well as reinforcing the benefits of the network technologies that you receive.

- Dependence on a key person

  In a project team of today, there are not many spare resources. Usually, there will be one person on the team that expresses great interest in the technology. They will start learning the software and systems on their own. The rest of the team thinks that this is great. The team now depends on this one person. Next, the person leaves and no one picks up the slack since it is a support task in the project. The web page or whatever technology then falls apart and is not supported.

- Unstable technology

  This is not typically due to the software, but to the hardware and network environment in your organization. With a network that has a high error rate and high traffic, data can be lost. Response time can be slow. After being frustrated several times because the network is slow or unavailable, people often decide they will revert back to old ways. They will start faxing, calling, and visiting. Once this pattern is established, it is hard to break—even when the technology and systems problems are repaired. This is why we devoted attention in this chapter to all of the support elements to make the network stable.

- Nonintegrated technology

  We have all had the following experience. You purchase some electronic item and bring it home. After opening the box, you find that you need some specific component that was not included. Often, the missing part is to link two things together. Back to the store you go. If they don't have the item, the thing you bought may sit unused for weeks. This is an example of the integration issue. Normally, you will acquire software and hardware from several vendors. When you consult their manuals, they hardly ever refer to each other's products. You are left in the dark as to how you will integrate them. If you can't figure it out, you will end up taking information from one system and rekeying it into another system. The cost of lack of integration is more labor efforts as workarounds. The net effect is to reduce the productivity benefits and discourage casual users of the technology.

- Making the wrong technology choice

  All of us have done this. With extensive data collection and analysis, we can still pick the wrong technology. You probably have neighbors with obsolete automobiles or appliances. It is the same here. Your organization may have picked the wrong electronic mail system. How do you know it is wrong? Typically, it is not being supported. There are no new releases. When you go to a bookstore, there are only out-of-date books on the software title.

What is the impact of the wrong technology? People are reluctant to use it. They think they are employing a dinosaur. The software lacks features of newer products. There are no interfaces to current releases of software. The net effect is to lower confidence and to increase labor effort to circumvent the limitations. What should you do? Step back and get a wider perspective. It is time to design an overall architecture that identifies what your long-term solution will be.

EXAMPLE: SHARED PROJECT INFORMATION

What occurs when people share information? There is a change in roles because people do not have to go to the project leader to get information. It is available on-line through the network. Since it is visible, it tends to be more accurate. People take more care with the quality of the information. They are also less likely to maintain their own information if it overlaps what is on the network.

In project management, it can happen that each person on a team might maintain their own schedule. They do this because the general schedule does not contain the detail that they require. This was common with older project management software systems which had less capacity and were slower in operation. The newer releases of the software can support larger schedules. If each person has their own schedule, you can imagine the nightmare of updating the overall schedule. You have to get information from each person and then interpret what the tasks and milestones mean (since they are not the same as the overall schedule). Then you must post the schedule updates. By the time you do this, it is time to start the update process all over again.

In a shared environment, each person works off of the same summary schedule. They are assigned specific task areas in the schedule. Each can go into their area and insert more detailed tasks. They also access the schedule to update the work. The project manager does not update the schedule. Instead, the project manager reviews the updates of the team for consistency with issues and action items. The obvious benefit is reduction of duplication of effort. There is also less time spent in coordination and obtaining consistent information. By working together, people tend to share information and lessons learned. These benefits can, in turn, increase morale.

EXAMPLE: PROJECT WEB PAGES

With newer software, we can produce charts, tables, and text in HTML and other Web-oriented formats. These pages can then be integrated and composed for the Web. The Web can either be on an internal intranet or the external Internet. The evident benefit is that information can be posted quickly and graphically for easy access through a Web browser. All of that is good. On the other hand, if this is not planned, the Web page may be established one time and never updated. Multiple projects may establish their own pages which are totally inconsistent, making it difficult to examine multiple different Web pages of various projects. In addition to network support, Web pages require: central coordination and assistance for project members to call, guidelines for development and maintenance of the Web page, and a sample page. The coordinator of the Web pages must monitor the pages and ensure that each project keeps its Web page updated.

EXAMPLE: LESSONS LEARNED

Tracking and using lessons learned is a living, dynamic function. It is more than just gathering up lessons learned, categorizing them, and posting them on a system. To be useful, the following steps should be taken:

- Identify a candidate lesson learned
- Analyze it with reference to all projects and the existing body of lessons learned
- Try to extend the lesson learned and determine impacts and benefits
- Identify the necessary policies, procedures, and systems steps necessary to have the lesson learned applied in projects
- Establish a monitoring process to see how lessons learned are applied

Management must give credit to staff for developing and using lessons learned. There should be a dual reward method where some benefit is provided to the submitter of a key lesson learned and to the person who got the most out of lessons learned.

Lessons learned are not static. As people apply one lesson, they gain more insight and experience and add on to the lesson learned. They also may have comments as to when the lesson should not be used. Thus, associated with each lesson learned, there should be a log of comments of people who have worked with it as well as impacts, examples, action items, etc.

EXERCISES

The first step is for you to identify and assess the quality, compatibility, and performance of the network technology in your organization. Focus mainly on the software tools. Next, identify what could be done to improve the technology.

From the answer to the above question, determine how project management could benefit from the current software tools. What process changes would be required to take full advantage of the technology?

Next, assume that you had access to all of the software tools that you specified. What process could you put in place? How is it different from the one identified in the previous answer?

EXAMPLE

The initial tools available were shared project files on a file server and electronic mail. This allowed communication across shifts without paper and shared viewing. The existing wide area network was expanded for additional workstations. After all schedules were established on the network, then the staff were trained in project management and electronic mail software. Sample mail messages were provided as examples. Project templates were set up along with project schedule files for current projects. Project managers updated the schedules; team members accessed and viewed the schedules for a month. After this, team members were required to make their parts of the schedules more detailed as well as to update their schedules. The next step was to have the project managers meet to discuss resource allocation over the period of the month following the meeting date. Management then worked with multiple projects by combining the schedules. This was possible due to the use of the standardized templates.

SUMMARY

Adding network technology and tools to project management can have a dramatic impact on productivity, control, and participation—if the project management process in the organization is modernized to take full advantage of the technology. What was done in a standalone, discrete, or batch environment moves into an on-line, participatory environment. With project templates management can review multiple projects in a compatible format. With shared project access, all of the project team can work from one copy of the schedule on the network without having to have their own schedules.

III MANAGING THE PROJECT

9 Tracking and Monitoring the Project

Let's recap where we are. In Chapter 4 we set up the general process for project management. We addressed the details in Chapter 5. We got our manager and team together in Chapters 6 and 7, respectively. How to use modern network based technology was covered in the preceding chapter. The project is underway. This chapter will consider what it takes to monitor, track, and control the active project. We will also discuss how problems are generally handled.

Why do we want to track and monitor the project? To know where it is going. But also, to be able to take advantage of potential improvements and to resolve problems and questions early so that they do not impact the project.

We can also consider tracking from different perspectives. Five perspectives and typical objectives are: project team (work satisfaction), end user of the project (end results), line management (resource use), general

151

management (cost and schedule), and the methods or tools used (effectiveness of use).

WHAT IS HAPPENING IN THE PROJECT?

The people are working on the project. As in many cases, there are varying levels of involvement depending on the specific team members. You may or may not be receiving questions or comments about specific administrative issues as well as technical project issues. The project may be being worked on at several locations.

The role of the project team member is to complete the assigned tasks. The member should also alert the manager about any problems or issues related to the project or to the specific tasks. The team member should take the initiative to approach the project manager with issues and opportunities. What types of issues can arise while project work is in process? Tasks may take more or less time than allotted. New methods for accomplishing project tasks may be conceived. There may be additional tasks that were not foreseen earlier when the plan was approved. In doing a task, you may uncover additional information that changes the nature of the task. Another possibility is the need for reworking and redoing tasks to address specific technical problems.

The project manager must act as a manager and not just a project administrator. In many projects, the effective manager participates in doing the work and dealing with tactical issues. Because of the importance of the management aspects of the project, we will focus on these first.

UNDERSTANDING VERSUS DECISION MAKING

We track a project because we want to assure ourselves that project objectives and strategy are being achieved. We also want to resolve problems before they become bottlenecks. We have two interrelated, and sometimes conflicting activities. One is to gain an understanding of the project, and the other is to facilitate decision making. We pass through stages in terms of observing without acting, planning to decide, making a decision, planning actions, taking actions, and following up on the actions. If we rush through the process, we would do all of these things in one meeting. In project management, this is a mistake. Off-the-wall actions can tear a project apart.

ISSUES AND OPPORTUNITIES

The identification, tracking, and resolution of issues and opportunities that could impact the project are fundamental activities in project management. They overlap administration and management. An issue is a potential problem, situation, or factor that can negatively impact the project. An opportunity is exactly that—a chance to change something in the project and achieve benefit. Issues and opportunities typically relate to resources, structure of the project plan, methods, and tools. Modifications here impact the schedule, risk, and cost. The source of the issue or opportunity may not be internal to the project; it can be external in terms of management, technology, another company organization, external competition or industry, government regulation, or another project.

Issues form like clouds that can hover over a project for a long time. Some issues can appear to be resolved and then return in different forms. People on the project might want to have the issue revisited later if the first resolution was not acceptable. When we discuss tools in Part IV, we will discuss the issues and opportunities database that can be either an automated or manual method for tracking issues.

It is appropriate to ask why you would not want to take advantage of an opportunity. The question boils down to trade-offs. If you seize the opportunity, you may disrupt the project. Enough opportunities can shred a project, so unless there is substantial benefit with very little risk, you should be conservative in your approach and the endorsement of opportunities.

A further point on opportunities is that it is often advisable to take advantage of several at one time in order to minimize the impact and disruption of the project. This applies to methods or tools. New techniques should be compatible and not counterproductive in terms of existing methods and among the existing techniques. Bear in mind that implementing something new requires control and support. If you cannot support something new and verify that it is being implemented properly, then, perhaps, you should not use it; if it is in use, then you might abandon it. However, once abandoned a method or tool is difficult to resurrect.

STEPS IN ADDRESSING AN ISSUE OR OPPORTUNITY

Let's briefly consider the steps in addressing an issue because we are going to be referring to this quite a bit later.

1. Understand the issue or opportunity itself and distinguish it from symptoms or impacts or effects. This step requires thinking about the project itself as well as the specific issue or opportunity. The issue or opportunity should be expressed or framed in the context of the project.

2. Understand the ramifications of taking potential actions or of inaction. Here we have defined the issue or opportunity and are attempting to understand how the project will be impacted by doing nothing or taking some action. You should ask what would happen if you take action but it is delayed.

3. Decide on a plan of action and see if by changing the action in terms of scope, you can resolve other issues. The idea here is that when you take

action, it may be a change to the schedule and, hence, an impact on the team. If you are going to be making changes, then why not bundle in several at one time?

MANAGEMENT ACTIVITIES

The following activities will be considered here:

- Determining the status of work on the project
- Analyzing the state of the project in terms of the schedule
- Assessing the quality of the work
- Motivating and encouraging the team
- Getting issues and opportunities resolved
- Reporting to management

Here we consider the content of what is to be presented: Chapter 14 discusses the form.

The project manager must "keep on top of the project" at all times. Even areas and activities that are either minor or seemed to be making progress last month or last week need to be checked.

COLLECTING INFORMATION FROM THE PROJECT TEAM

When we collect information on the status of tasks, we are also collecting information on all parts of the project. Project managers should make an effort to have a unified and integrated approach to data collection. That is, we should collect data from a team member at one time for all topics as opposed to going back to the person several times. The latter approach consumes time and reduces productivity. You should collect information from a team member in their office or work area, not yours. Before going to visit them, sit down and go over what you know about their tasks. Review the list of active issues and opportunities. Do these apply to the team member's work? If their work is not on the critical path, then how close are they to the critical path?

When you are in their office, let them talk about their tasks in their own words. Do not start with a checklist of all their tasks and go down each task. This will inhibit them. Just say that you are preparing the normal progress report for the project and want to know how they are doing. Let them talk about the work in general. Then after they have talked, go over any relevant issues or opportunities. Sharing information may elicit suggestions on their part and they will feel closer and more involved in the project. After you have discussed the past and current situation, ask them about the work coming up over the next few weeks. As you are having this conversation, don't hesitate to keep notes. You should also encourage them to tell you what else they are working on.

At the end of the meeting, you should summarize the status of the project in terms of what the team member is working on so that there is no misunderstanding. This gives them a chance to express any opinions or make any corrections.

COLLECTING INFORMATION FROM OTHER SOURCES

You will be attending various meetings as a project manager. Add in meetings with contractors, subcontractors, people from other projects, and others from whom you may pick some other information. When you finish a meeting, sit down and ask yourself immediately afterward what the potential impact is on your project, if any. Are there resource issues? Is management adopting a new method or tool that will have to be used on projects? Is another project in trouble so that they may need to take resources from your project? Also, ask yourself what their body language, tone of voice, and feelings told you about the project.

COMMUNICATING WITH NETWORK TECHNOLOGY

With projects spread across the world or with many team members, in person and telephone contact will not suffice. Electronic mail is an obvious tool for obtaining status on projects. It is—if you follow some guidelines.

- Use a standard format for the message
- Status should cover the following:
  - Status on active tasks
  - Additions and changes to the schedule
  - Identification and discussion of any issues
  - Projected schedule for the next reporting period

After receipt, the project manager should review it carefully and ask questions to verify that the information is complete and correct. As an alternative to electronic mail, a staff member can fill out a form in word processing and save it as an HTML (Hypertext Markup Language) file and post it on a Web server. With encryption the Internet can be employed as well.

Groupware is a good tool for discussing issues and building lessons learned. Groupware allows for a specific form and joint access. If you use groupware and go to the effort to develop groupware templates, then you must follow up with rules to ensure use.

You should consider allowing access for line managers and others. They can review and monitor the project. There is some exposure here, but we feel that it is offset by the elimination of misunderstandings and miscommunications through information sharing.

PROJECT ANALYSIS

You have information from some of the team members or from outside sources. Should you wait to do the analysis until you have input from all team members? No. Start right after your first update meeting with staff. You can start building a project update any time as soon as you get information.

How should you proceed with the update? First, you will be doing it bottom up. Go to a specific task and make a note next to the task as to the status. Start with the tasks on the management critical path. Go to each issue or opportunity and add notes as well. Now step back and take a wider perspective. How is the project different now from a week or two weeks ago? Has there really been progress? It is often easy to get caught up in the detail and the trees, and miss the forest. This is also important when you have to meet with the project team and review the status as was covered in Chapter 6. It would be nice to report specific progress as opposed to percentage complete. In collaborative scheduling each team member can access the schedule on the network and update their own tasks as well as adding additional detail, rework, etc. This function works well in conjunction with using electronic mail, the Web, and groupware.

What happens if there are no major milestones being achieved? Don't panic. It could be that the project is in a calm state. But if it goes on week after week, do not assume that things are OK. The first place to look is the plan and the list of tasks and milestones. Did you include sufficient milestones so that you could track progress? If not, then you might revise the schedule. If this is not the problem, the you should consider the work itself. What is on the management critical path and how are these tasks going? Go to the other extreme and look at the tasks that have the greatest slack and are furthest removed from the critical path; how are they doing?

Look at the infrastructure of the project. This includes the project structure, methods, and tools associated with the project. How do you review the infrastructure? You should think about potential changes to move tasks from dependency into independence and parallelism. How are the tools being used? Is there make-work effort going on just to satisfy the requirements or procedures? Are people spending time on useful work that is not part of the tasks? If so, consider adding tasks.

DETERMINING PROJECT STATUS

Having updated the detail, you can start to look at project status. If you use a computer-based project management system, then you could update the tasks that have been worked on by the amount of effort. You could also change the status of any completed milestone to completed from active or in process. In doing the update either manually or using automation, start the update with the mathematical critical path. Rerun the schedule and compare it to the planned schedule. Look at any changes that occurred. Did any new problem surface? Then update the managerial critical path and rerun the schedule. When you review the comparison of actual versus planned, look at future milestones and see if they have shifted. If they have, then you can look next at tasks leading up to the milestones. If they have not shifted, then consider task differences based on specific resources. Follow these steps:

1. Enter updated information.
2. Review the new mathematical and managerial critical path versus that of the previous schedule.
3. Review planned versus actual overall schedule.
4. Review noncritical active tasks.

Detailed schedule analysis is accompanied by an assessment of issues and opportunities. Since the last update and review, has there been significant progress in the resolution of issues? Do we have a better understanding of the issues? If the major (highest ranking or highest priority) issues do not show progress, then you may have a substantial and growing level of risk in the project.

Because a project manager cannot focus on everything, you have to determine in this specific review if the issues or the schedule dominate what is important. The following cases often occur. Note that issues here include opportunities and issues. Limited progress means either increased problems or very little work done.

1. Issue: limited progress; schedule-limited progress. Here you would start with the schedule and determine if the reason for lack of progress was that some of the issues were impeding progress.
2. Issue: Progress; schedule-limited progress. In this case you would review the schedule and determine if there are new issues that need to be addressed. Focus is on the schedule.
3. Issue: limited progress; schedule progress. Progress in the work will not continue if the issues do not get resolved. Focus on the issues.
4. Issues: Progress; schedule progress. The tendency might be to just report on the schedule and focus on it. This is wrong—move to accelerate work on the issues.

Notice that in most cases, we suggest that you give your attention to issues. Why? Issues are where the major project risk is. Allow the project team to do the work and contribute to identifying issues and opportunities. The responsibility for getting the issues resolved in the project lies with the project manager.

PROJECT UPDATE

We have updated the schedule and the issues status or database. We have analyzed the updated schedule. We are ready to present the status to management and to the project team. Although the management presentation will have an introduction of schedule versus planned and actual versus budgeted, the attention in both settings should be on the issues and any schedule problems. Do not spend too much time giving a detailed status report on tasks that are on track and on time. Focus on the problem areas and what you are going to be doing about them. You will have to be prepared to answer questions and issues related to any project concerns. Table 9.1 gives examples of two status reports—one focusing on issues and one on schedule.

EXTERNAL PERCEPTIONS VERSUS REALITY

The world outside of the project will pick up impressions of your project. While appearing harmless and not part of the actual work on the project, these impressions may impact the project. If people perceive that a project is in trouble, the project team may start getting pulled apart. Managers may not want to have their staff spend time on a failed or troubled project.

Another impact is that it may be more difficult to get management approval or even management attention. After all, who wants to hear bad news? So, the message that is conveyed from the project to the outside world is important. The first suggestion we have is to ensure that the true, factual state of the project is known to management. It does no good to hide problems. Get them out on the table and address them. Make sure that all of the project team is aware of all current issues and their status.

TABLE 9.1   Examples of Two Types of Status Reports

This project is to install a new financial system on a computer. The specific system that is being considered here is the general ledger subsystem.
A. Status-oriented
   1. Worked on definition of chart of accounts for general ledger. Resolved two levels of accounts.
   2. Reviewed training requirements for accounting staff for new system. May need to train more people than originally estimated.
   3. Began the setup of test data for the new system.
B. Issue-oriented
   1. Level 3 chart of account definition still open. Disagreement between auditors and general accounting on structure. Hope to resolve next week.
   2. Training for staff now 40% more than planned. Working with training department to resolve.
   3. Test data being set up—no problems.

Keep people updated through one-on-one contacts as well as through project meetings. In other words, an open process is superior to a closely held approach where there is minimal communications.

QUALITY ASSESSMENT

We have indicated that you will be discovering the status of the project and milestones. There are levels of determining status. You can ask someone if something is done. Then the answer you receive is not supported by evidence. Next, you can ask for some tangible sign that the work has been completed. This could be a report or some physical structure. Seeing it would be a review by presence. You have no idea if the work is any good. A third level of review is for assessment of quality, content, completeness, and other attributes. This means that you have to get inside the milestone and see the details. This takes a substantial effort. Usually, you have to be very selective as to what milestones you will review in detail. Reviewing too many will mean delays in the project while people wait to hear from your review.

How to Review Project Milestones

First, you will have to decide on the level of review needed. You have some flexibility here. You could start at one level, such as presence, and then if you uncover problems, you can expand to the next level. This leads us to our first recommendation here—adopt a level or tiered approach. That is, define a series of levels and an escalation process to go with it.

You need to define your approach and resources in the review. Who has the knowledge of the technology or situation as well as that of the project? These people are hard to find. There is another question that we raised earlier. What is to be done in the project while the review is going forward? Do people continue to work on the project, or do they stop and work on other projects while the review is going on? This leads to our second recommendation—timely and well thought-out reviews. If you stop or disrupt the project during the review, the project loses momentum and it is difficult to get the project moving again. Some suggestions are

- Plan ahead for what people will be doing during the review;
- Determine what your approach will be for the review in advance and line up reviewers.

In doing the review, you should follow a two-step approach. After the review starts, ask for feedback early to determine the general scope of any problems. This will give you information on what you should do in the project if immediate action is needed. For example, if the design of a braking system for an automobile was being reviewed, you would like an initial review in terms of safety. Then you would ask for a more detailed assessment on the quality of the design. The second step is the detailed assessment of quality or work.

The Results of the Review

The review may indicate several things. At the extremes are failure and endorsement of the milestone. These are unusual. Usually, the findings are mixed with shades of grey. There are suggestions for improvements. There are points where the reviewers did not understand what the builders were doing. What do you do? This is an excellent time to reassess the project as discussed when we covered status. This might be the time to make several changes in the project that improve the project infrastructure as well as take advantage of the findings. For specific findings you have to decide what action to take. You can stop the project, reorient the project, add to or slightly modify the project, leave the project alone, or take other measures. You will get one chance, because once you have decided and announced what is to be done, the die is cast. Work will resume on the project.

Making Changes in the Project

We have pointed out numerous cases where changes can occur in a project. These changes tend to be loaded toward the beginning of the project and around areas of risk. But they can occur at the end of a project in a desperate attention to salvage something from the project. You can and should make several changes at one time. How do you make changes? First, having decided on the changes to make, you would identify what is to be done differently. Second, you would turn your attention to how the work is to be done. This includes the methods and tools. The third step is to reassess the project infrastructure in terms of project structure and organization.

After deciding on the changes, the next step is to define your expectations for these changes. In other words, you are not making the project changes for fun, but to achieve a specific effect. What is the effect? Next and related to effect is the question of how you will be sure that the effect or impact has been achieved. What are the signs of success?

Following up on the effect and signs of its occurrence is control. How will you be sure that the changes are being followed? How will you know that the project is not just going on like before? This relates to project control, which we will discuss later.

The Impacts and Side Effects of Change

What are the impacts of change? Aside from lost time for reorientation and regrouping, people can be refreshed or demoralized. The follow up will be important.

If changes are made and the project structure remains the same, the impression is sometimes given that there has been no real change. After all, if there were real change, then how could the project task and milestone structure remain intact? Thus, the project plan needs to be changed. When the plan is presented to the project team, the changes should be explained in terms of how the new structure relates to the changes in the project. This applies to resources, methods, approach, and tools.

─────── PROJECT ADMINISTRATION

Project administration includes documenting the results of reviews and keeping track of information about the project generally. The major point here is organization and structure. You as the project manager need to be able to quickly get your hands on almost everything about the project. This is not just for management questions; it is also to respond to issues and questions raised by the project team on a specific issue. The project manager is the basic source of knowledge and experience on the project. In large projects, this is called the project office. In the "luxurious days" of the 1960s, many projects had project administrators as well as managers. Well, as Chapter 1 said, today is different; we have to do more ourselves. Organization means collecting the information, storing it for ready access, and being able to retrieve and anlayze it quickly.

After updating the schedule and issues, you can publish the results on the Internet Web site for team members to access. This short cuts faxes, mailing, telephone tag, and generating and distributing paper. Be sure to indicate the date of the information and highlight any new issues and changes to encourage dialogue among the team.

PROJECT CONTROL

Project control here means the process and steps taken to control the project at the project level. Management control is the review and control of multiple projects at the managerial level, by way of contrast. We have already outlined the process for tracking the work. Here we want to give attention to the consumption of resources and costs.

Looking at a project budget we find that the human resources and overhead are basically fixed. In some projects we can watch overtime if that is appropriate. However, it is often the expenses that can raise the red flag. These expenses include contractors, subcontractors, travel, equipment, training, and supplies. There is a danger in the project of consuming the monies allocated for these items too early in the project and not having any funds left over for contingency or emergency.

In terms of control we not only need a total amount for each budget category, we also need a rate of expenditure that is realistic. How to estimate the run rate or rate of expenditures is usually based on experience. When in doubt, estimate bottom up. That is, identify each item and then attempt to place it as far out in the future in the project as possible. Beware that when you estimate bottom up, you tend to round up resources and costs. This leads to a total number that is unrealistic. What to do? After estimating bottom up, then revise top down.

Because many projects depend on fixed resources, the normal budget review may not make sense. It might be better to look at how staff are allocating time to the project. A question here is whether staff should allocate time between specific tasks. This is controversial. Allocating time sounds like a good idea. It would provide useful information. However, it means the added time and expense of having employees fill in this information and then having someone tabulate the information. Even then, there can be the danger of employees just putting down hours to match the plan. For example, if the plan calls generally that you will spend 40% of your time on A and 60% on B, then you would allocate 16 hours to A and 24 hours to B in a week. So, for all of this effort, we get exactly nothing but the plan back in our face.

SIGNS OF PROBLEMS

How can you tell if there are signs of problems in the process of project management? There are the obvious signs of a project in trouble in terms of budgets, delayed milestones, and so forth. But these often appear after the fact. There are signs to look for in the project itself. Is there good communications between project team members and the project manager? A lack of communications means that the project is in a higher state of risk. You just don't know what is really going on.

Another problem sign is vagueness in the status reports and information. There are pleasant words of progress and small issues being overcome, but major issues are not being addressed and no significant milestone has been achieved in some time.

A third problem is attitude of people. You can tell if there are problems often by the attitude of the staff. People who talk in low voices and look down at the floor repeatedly are people who know something is in trouble. Other signs are attempts to jump ship and move to another project.

DIFFERENT POINTS OF FOCUS DURING THE PROJECT

We have treated the project the same regardless of the stage of the project. Although many of the processes and situations are the same, there

Example **163**

are some differences. At the beginning of the project, there is a certain fluidity. This is a time when patterns in project management are being set. The team is getting used to working together. The actions, management style, and controls employed can impact the entire life of the project.

Typically, after this initial period things may settle down in the project until the first crisis occurs or major issue surfaces. Then comes the test of what the project manager will do and what role the team will assume. Will they let the manager address the issue alone? Will they provide active support and involvement? How the manager and team deal with crisis will set the pattern for later issues. After this there are likely to be changes at the time between the ending of a phase and the start of the next phase. If the project was started off right, then these should be routine. Thus, even if there are more severe crises later, the project team can weather the storm. Usually, the project experiences peaks and troughs of crises over time.

EXERCISES

Take a specific project you are or were involved with in the past. Identify how project status was obtained. What happened at project review meetings? How were issues identified and addressed? Was there an open process that encouraged people to come forward with issues or questions, or was it a closed process where people did not want to hear about problems? These are some of the questions you might want to ask.

EXAMPLE

We will look at several aspects of the project that pertain to the monitoring and control of the project. First, we indicated that the initial project work included a series of meetings with the areas of integration and final assembly. These identified over 100 issues (many of which were not within the scope of the project, but pertained to specific technical issues). This was useful in getting a list of issues that could later be tracked and monitored in terms of progress.

In terms of review, it was relatively easy to conduct reviews within the core project team. What was more difficult was to conduct informal, ad hoc reviews with people in the areas and in line organizations to see what they thought of the progress and issues in the project. Perceptions are important. If people perceive an issue and do not see progress in getting it resolved, questions arise regarding the project and the team.

The managerial critical path was entirely different from the mathematical critical path in this project. The mathematical critical path was generally composed of technical tasks related to client server software development, implementation of the wide and local area network, testing, and documentation. If we had kept our eyes on this, we would have failed. The real critical path was identifying and addressing issues involved in the reengineering of work. Logistics and getting the new work flow in place were also critical. Demonstration of design and prototypes were also critical. But because these tasks were short and separated, they did not appear on the managerial critical path. Some upper-level managers recognized this and asked for status on only the managerial critical path. One called it the political critical path.

The project underwent changes. When the project started, the approach was to implement the new work flow and project management process across all areas at one time. This did not work. There was too much detail and a lack of focus. After a regrouping planning session by the core team, the approach was changed to focus on the first areas where final assembly was performed. Integration and testing were then emphasized. The effort in reconciling the project plan to reality required tasks to be modified and added as well.

Another issue that occurred in the project was that of getting participation by area managers. Individual review and work sessions proved less effective than group sessions with several areas at one time. This provided encouragement for participation.

A valuable tool in getting the schedule to be accurate proved to be ensuring project data visibility to managers and staff. This immediately highlighted any area where there was less than adequate participation.

As was stated earlier, a list of critical resources was used in the plan. An interesting behavioral change occurred as a result of this. When an area requested money for additional equipment, the other areas insisted that the area show the benefit of the additional resource. This actually saved several hundred thousand dollars in requests.

SUMMARY

In summarizing the chapter we first emphasize that reviewing and monitoring a project is a continuous process. We constantly are collecting information on the project. At times we use the information to produce an update to the project plan and report to management. At other times we use the information to address specific issues and opportunities. Issues, opportunities, and the events in the project lead us to potential changes in the project. We have pointed out the need and benefit for discrete changes to the project to ensure stability. We also have emphasized that the project plan and structure must mirror and reflect the changes in the project. Otherwise, the changes lose credibility.

Note that our example discussion above has changed to include some results and guidelines in applying what we have discussed. This will now continue in later chapters.

We have tended to focus on issues as opposed to the schedule itself. There are several reasons for this. First, issues and opportunities and their resolution lie with the project manager. The tasks in the project are performed by the project team. The manager could also work on some tasks. Thus, it is appropriate for the project manager to give attention to the issues. A second reason is that risk will generally relate to the issues. Taking care of the issues will tend to minimize project risk.

10 Project Cost Analysis

It seems simple—you assign a resource to a task. A resource has an assigned cost. As the resource performs the task, a cost is incurred. You compare what you planned for it to cost with what it actually cost—that is, budget versus actual. It's more complex than that.

Return to the gardening example and assume that the entire project is one task. In terms of cost, we have to buy plants, seeds, and fertilizer at the start. If you rent a cultivator to break up the ground, then you have a rental fee at the end of the project. If the project takes longer than planned, the rental fee will be more than planned. Let's suppose that you hire someone to do the ground preparation and planting. That person works at an hourly rate, may get overtime, and is paid by the day.

This simple example reveals some of the complexity in costs related to a project. You can have resources that accrue costs:

- At the start of the task. These are typically resources for which payment is required in advance.
- As the task progresses (progress relates in a linear or proportional way to time). This is common for analysis, but can lead to errors since costs are applied discretely in an accounting system.
- At the end of the task. This is the most common. It does not apply to a detailed task that is very long.
- On a per use basis (the cultivator has a flat rate for a week minimum, but we only need it for two days). This is common for equipment and facilities.
- Combinations of the above. These make it even more difficult for a software system to accommodate variable costs.

For each resource you must first determine if it is to be assigned a cost and second how it is to be assigned and accrued.

There are some other factors involving costs:

- Overtime and double time for holidays and Sundays. This can get complex.
- Union rules on computing pay and costs.
- Calendar that is employed for the resource and task (number of hours per day, per week).
- Overhead costs.
- Supervisory and oversight costs.

Most of these are difficult to handle within a project management software system. This is not unexpected since a project management system is not a timekeeping or accounting system.

STEPS IN CALCULATING PLANNED COSTS

You have several options for handling costs inside or outside the project management software. We will consider both in our first part, calculating planned cost. The basis for the steps is that most people and computer software programs calculate work based on resource assignment to a task. The total work for a task is the sum of the work performed by all of the resources assigned to the task. However, total work makes little sense if the resources cost different amounts of money. To generate the cost of a resource in a period of time such as a week, you would add up all of the work performed by the resource across all resources. If the total number of hours worked exceeds what is available according to the calendar for the resource, then you have overallocated the resource. Resource leveling would then shift the task's start and end dates to eliminate the overallocation.

Note that if you don't assign resources to a task, you cannot calculate work since it does not exist. You cannot do resource leveling or determine any resource conflicts. In short, you then are using the project management software as a fancy drawing tool. You can only do very restricted what if analysis—changing only durations and dependencies.

Here are the steps in determining the cost of the plan from the schedule.

- Step 1: Set the baseline plan or schedule. This means that you start with the project template and add detailed tasks, more dependencies, resource assignments, and durations to generate the plan. You then do "what if" analysis and make changes. When you are satisfied, you save this as the baseline plan.
- Step 2: If you are going to assign costs within the project management software, associate a cost and accrual method for each resource that has a cost. Most systems allow for only regular and overtime or per use costs. Accrual is at the start, at the end, or as worked. If you don't assign costs here, go to step 3.
- Step 3: Obtain the table where the rows are resources and the columns are periods of time (e.g., weeks, months). The table entry is the number of units of time that the resource worked. This is sometimes called a work usage table. This table is normally a menu choice in the software.
- Step 4: If you have assigned costs, then generate the table of cost usage. The rows are resources; the columns are units of time; the entries are costs. This table is derived from the work usage table in step 3. If you don't assign costs, go to step 5.
- Step 5: If you did not assign costs, export the work usage table to a spreadsheet. Once in the spreadsheet, you can assign costs and do calculations. If you did assign costs, go to step 6.
- Step 6: If you assigned costs, export the cost usage table to a spreadsheet. You can now refine costs using the spreadsheet.

TRADE-OFFS—ACCURACY VERSUS EFFORT

Will the steps yield accurate results? You might answer "yes" since the schedule is correct. However, it is normally an approximation since you usually have not included every resource and you have not included overhead, supervisory time, union rules, etc. You have already seen that these are difficult to include in the schedule.

Detail is a trade-off. The more detail you add, the more accurate the schedule is. However, the more detail you add, the higher the burden you will pay in maintaining the schedule. A fundamental lesson learned is that you must generate a schedule that is maintainable and capable of being updated with reasonable effort. If you have a schedule for a one-year project that has 1000 tasks with each task having 10 resources, you will spend all of your time updating the schedule. Keep to less than 500 tasks with at most five resources assigned.

There is a dilemma here. In general, because the costs do not exactly match what is happening with the schedule, you are stuck with integrating a budget and a planned schedule. This is a basic challenge in scheduling. It gets more complex if you are estimating. Suppose that you have several schedules for projects and that you have a resource pool. You want to know how many units you can build with the resource pool. This is not automatic as there is no icon or button that does this automatically. It is an iterative process where you go back and forth. You determine what resources are required for each project. Then you start depleting the resource pool. You then move back to the number of units and make adjustments and recalculate the resources required. Continue as is.

GUIDELINES FOR BUILDING A BUDGET FOR A PROJECT

The above discussion reveals that guidelines are useful in building a budget. Here are some guidelines in the form of steps.

- Step 1: Construct the planned schedule and set the baseline.
- Step 2: Calculate the direct costs of the resources that are consumed in the baseline. This can be done in either the spreadsheet or project management software.
- Step 3: Now add overhead and other costs to the budget based on the rules of your organization. This is your first cut budget.
- Step 4: To refine the budget, go back to the plan and change the schedule or resource loading on the tasks. Repeat steps 2 and 3.

You should become proficient at carrying out these steps. Why? Management will typically want to adjust the budget and do what if analysis with the money. It is up to you as the project manager to return to the schedule and determine the impact of budget changes on the schedule. This is a major cause of problems with budgeting. People do not have a resource-loaded schedule and so perform what if analysis by seat of the pants methods. Do you wonder why so many projects later run over budget?

ACTUAL COSTS: PROJECT MANAGEMENT

Finally, you can start the project. After a project has started, costs begin to accumulate. New factors enter the picture. You are getting information from two sources: the updated schedule and the accounting system that is logging the costs chargeable to the project. Reconciling these can be a nightmare. Let's start with some remarks on the schedule.

If you use percentage complete for tasks, then costs will begin to accumulate before a task is finished. An alternative is to have tasks in one of the following states: not started, started, and completed. Detailed tasks are either 0% or 100% complete. Summary tasks can be some percent complete based on the detailed tasks. For example, if summary task C is composed of two tasks of equal duration and effort in sequence, then when one is complete, our summary task C is 50% complete. There is a basic problem with percentage complete. In working and costing the software assumes that costs are rising in a linear or proportional manner over the duration of the task. This may not be realistic. The project management software has a field for the actual cost of a resource applied to a task and the actual cost of a task. Actual costs are based on actual work. This corresponds to the way that planned costs are generated. If the durations, resource levels, or other factors change, the work and costs change as well. When you update a schedule, you mark some tasks as complete. Actual costs are calculated for the task. You also start some other tasks. If costs are accrued as a percentage of the time, then actual costs begin to accrue here. Tasks that are active and continuing have their percentages of completion increased, so their actual costs increase.

ACTUAL COSTS: ACCOUNTING SYSTEMS

While the project is going on, bills and time cards are being submitted to various systems. Each input is associated with a project through a project code. There is then a system that rolls up the total costs assigned to each project. Let's suppose that you get this report. What are some of the issues and questions that arise?

- The labor costs are totaled so that it is very difficult to determine how many hours each person worked on the project. It is also difficult to determine the allocation of the time to the active tasks to which they were assigned. For example, suppose Sam White worked on your project for 24 hours in a week. He had tasks 100 and 102 assigned to him. How many hours did Sam spend on each of 100 and 102? We have no idea—either Sam did not have to post to individual tasks because the payroll system could not handle tasks, or because there is no system for recording that level of detail. If you say that you want this detail, think about it carefully. You would have to set up an on-line interface between the payroll system and the project management system.
- Vendors submit invoices to accounts payable for work performed on the project. By the time you review and approve the invoice and the other steps occur, the costs may get posted to the project one month later. Accounting and project control systems tend to be lagging systems. This makes comparing budget and actual cost much more challenging.

BUDGET VERSUS ACTUAL COST

You now have the planned cost and the actual cost. These can be compared in terms of budget versus actual. However, you could compare any two sets of dates for a schedule. For example, you could have the customer contract dates, the dates planned by management, the planned dates by the project manager, and the actual dates of work. You could compare any two of these.

- How can the project be behind schedule? This is obvious. The actual completion date based on the actual dates is later than the planned completion date.
- How can the project be on or under budget and behind schedule? This occurs most frequently when you can't obtain the resources for the work. You are not incurring these costs; the work is not being done.
- How can the project be under budget and ahead of schedule? You could have overestimated the task durations. You are actually finishing the work faster than anticipated. Since there is less work in the actual task, there is a lower actual cost.
- How can the project be over budget and on schedule? Typically, the project manager applied additional resources or more hours of existing resources in the elapsed time to keep on schedule. The project is on time, but has an overrun.
- How can the project be over budget and behind schedule? Here you usually underestimate the work. The plan is optimistic. There can be unforeseen rework. New, unplanned tasks surfaced. Even if we apply more resources, the schedule might slip.

How can you graph or display actual versus budget? One example is shown in Figure 10.1, which reveals actual versus budget for cumulative costs. This figure is useful in getting an overall picture. In the example the actual costs underran the planned costs until recently, when the actual costs exceeded the planned costs.

For more detail, you can consider a histogram of costs per period. This shows when actual costs exceeded planned costs. An example appears in Figure 10.2. You can move from this chart to the GANTT chart to see in which tasks the problem occurred.

A third method is to consider costs or work by resource. The table below indicates the resources versus the actual and planned costs. This can pin down an issue by resource as opposed to task. It is particularly useful when there are many tasks and a limited number of resources. You can then go to the schedule and filter on the resource to see which tasks were performed during the specific period.

Resources	Budget	Actual	Variance
Resource 1	45	40	−5
Resource 2	50	55	+5
Resource 3	35	35	0

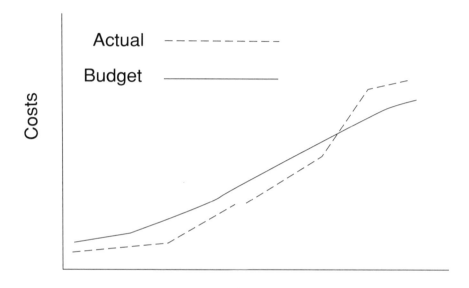

FIGURE 10.1 Actual versus budgeted cumulative project costs.

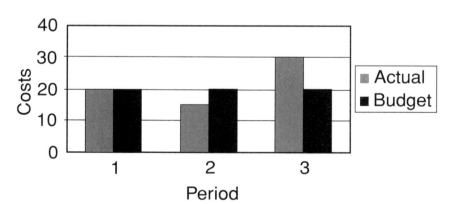

FIGURE 10.2 Actual versus budget by period.

EARNED VALUE

Earned value is based on the original estimate of the schedule and the progress to date to determine if you are on budget. Other terms synonymous with earned value are performance measurement and management by objectives. Let's take one task and work through a simple example. The task characteristics are:

Task A

- Planned or baseline duration—10 days
- Planned start date—May 1
- Planned end date—May 12
- Resources assigned—Person 1: $20/hour; Person 2: $10/hour
- Calendar—5 days/week; 8 hours per day

The total planned cost of the task is $2,400 (80 hours x ($20 + $10)). Now work begins. The actual start date is May 1. At the end of one week (May 5), the project is only 40% complete. Both resources have worked for the entire week. The actual cost incurred so far is $1,200. However, since you still have 60% of the work to go, our new estimate of completion is May 17, assuming that they are working at the same rate and there is no new slippage. This is a slip of 5 calendar days and $3\frac{1}{2}$ working days and an overrun of $600.00 for a new total estimated cost of $3,000.

Let's use this simple example to examine some of the concepts associated with earned value. Note that if you assigned no resources to a task, you have no work and no costs or earned value—another reason to always assign resources to tasks.

- Budgeted Cost of Work Scheduled (BCWS) is the earned value of the task based on the plan. BCWS is also called scheduled work. The planned percentage complete on May 5 is 50%. Multiply this percentage by the planned cost ($2,400) and you get $1,200 as the BCWS.
- Budgeted Cost of Work Performed (BCWP) is similar to BCWS except that it is based upon multiplying the percentage complete (40%) times the planned or baseline cost ($2,400). BCWP for our simple example is $960.00. BCWP is also called performed work.
- Scheduled Variance (SV) is the difference between the two above figures (BCWP and BCWS) or $240.00. If this were negative, we would be doing well since we would be beating our plan. Here we are not. We are in the hole.
- Actual Cost of Work Performed (ACWP) is the sum of all actual costs incurred to date. This includes any fixed costs incurred for the task. For you, this is $1,200. If we were to have people work overtime, the costs would be higher.
- Earned Value Cost Variance is the difference between what was planned (BCWP) and what occurred (ACWP). This tells if the task is on target

with reference to costs. In the example the number is zero so you are on target so far, but you are to suffer later since you project slippage.
- Total projected cost for the task is $3,000. This is also called Forecast at Completion.
- The cost variance is the difference between the baseline or planned cost and the actual cost. The variance is $600 and positive—you are going to overrun. If this were negative, you would underrun.

Many people like to use earned value to measure the budget or planned versus actual since it reflects costs as well as budget.

Let's consider another example. Suppose you have four departments or work areas that are performing tasks on a project. Suppose that their planned costs are as below.

Work area	1	2	3	4
Planned value	20	30	30	20

The total planned value is 100. Now the project starts. As work is done, it is earned on the same basis as planned. Planned value versus earned value measures the dollar volume of work planned versus the equivalent dollar volume of work accomplished. Any difference is a schedule variance. In the example below with sample data, the schedule variance shows that units 3 and 4 were completed. The schedule variance for units 1 and 2 indicate that the work is not completed.

Work area	*1*	*2*	*3*	*4*	*Total*
Planned value	20	30	30	20	100
Earned value	15	25	30	20	90
Schedule variance	−5	−5	0	0	−10%

You can compare earned value with actual cost as well in the next table. In areas 2 and 4 the earned value is less than the actual. More money was spent for work done than planned.

Work area	*1*	*2*	*3*	*4*	*Total*
Earned value	15	25	30	20	90
Actual Cost	10	30	30	25	95
Cost variance	+5	−5	0	−5	−5

ACTIVITY-BASED COSTING

In activity-based costing you divide up the project and organization into activities. With each activity we assign resources. Performing the tasks yields costs that can be priced. These prices can be assigned to task output. With each task, you can associate an average cost. Traditional accounting has been found to generate inaccurate costing information. Several reasons were the ability to determine total costs and limited management information for decision making. The total cost of the project or product is the total of all average costs associated with all of the activities. Based on experience you can update the average cost of future activities. You can also compare actual average costs with planned average costs. Idle time is reflected in the difference between resource use and resource spending. This method is often used for scheduling and planning. The limitation of this approach is that it assumes stability in the activities. Activities and tasks can change causing a problem with the average cost approach.

Activity-based costing provides the following benefits:

- The current cost of the task is captured.
- Attention goes to high cost tasks. However, these are not necessarily those with high risk.
- The method can support forecasting with average costs.
- Activity-based costing supports business process reengineering since reengineering redefines new activities.

If you embark on activity-based costing, you will follow these steps:

- Step 1: Assess the tasks to see what is to be costed.
- Step 2: Obtain the cost information.
- Step 3: Allocate the costs to the tasks.
- Step 4: Establish milestone or end-product measures.
- Step 5: Analyze the costs.

VALUE ENGINEERING AND KAIZEN COSTING

In value engineering you want to reduce costs associated with work. This is target costing. You examine how to reduce the cost of each task. Kaizen costing is the continuous effort to reduce and control costs. Both of these focus on costs. These are equivalent to considering the same approaches relative to task structure and to how the work is to be performed. There are some problems with this approach. One point is that attempting to reduce costs in the middle of the project can actually drive costs up by the disturbance you cause. Second, if you use the method, focus on the tasks that are in the near term future. This will be less disruptive.

Kaizen costing is part of the overall concept of "design for cost." This is not to be confused with design to cost. In design for cost you use technology and reengineering to reduce the project cost. In design to cost the project plan or design is refined until it meets a specific cost target. The goal here is simplification of tasks which in turn drives down costs.

VARIANCE ANALYSIS

Project management systems can generate a number of variance reports. You can filter these to restrict our attention to tasks for which the variance exceeds a certain level. In addition, you can design new reports from the fields in the project management database. Generation of the reports is the first step.

The next step is to access the schedule and examine those tasks with a variance. Check on the outstanding issues and action items for reasons for the variance. If it appears that nothing is being done, then you could generate a new issue.

REVIEWING A GROUP OF PROJECTS

Where should you start? First, you want to gather data on costs and schedules for all projects. You would want these to be in compatible formats for ease of analysis. Two approaches are:

- Start with cost reports and determine which projects are over budget. Consider to what extent the costs are up-to-date. Then move to analyze the tasks of the projects that are over budget. This is very difficult if there are no resources in the project plan and if the cost reports are not very detailed. All you could do then would be to ask people what is happening with the current work.
- Start with the project plan and determine which projects are behind schedule and over budget in terms of work hours. You can then analyze the tasks of the project in more detail. You could also move to the costs to see to what extent the project problems have hit the accounting system.

The second approach is preferred since it is closer to the work and does not have the lag time of the accounting systems.

PROJECT COSTING SOFTWARE

To address the issues of combining accounting and project management data, a number of project costing systems have been developed. These all share a high degree of integration between the system components. Components include: general ledger, payroll, timekeeping, accounts payable, purchasing, and expenses. Some of these have been developed in groupware such as Lotus Notes. What are some advantages and features of such systems?

- They can send automatic notification when a project falls behind schedule or is over budget by $x\%$.
- Payroll expenses can be automatically charged against a project or task area.

• Purchase orders to vendors can be tracked as can invoices against the purchase orders. Both can be charged to the projects.

EXERCISES

Ask yourself the following questions to assess the issues in costing out projects in your organization.

• How long does the accounting system lag behind the actual work in a project?
• How detailed are cost reports from the accounting system?
• Are resources assigned to tasks in the schedules?
• Is budget versus actual analysis based on the schedule or on costs?

EXAMPLE

In one business unit, there was a manager who wanted all of the detail in the project plan. He wanted the costs to match. He applied 10–15 resources for each task. Many detailed tasks were of one or two days duration. The manager went insane trying to update the schedule every day. He found it took two hours every day to update and change the resource mix and tasks. Even then, his costs were 20% off from those of the accounting systems. People who followed more standard processes were off by 30%—not bad for much less effort. You might think that this is bad, but consider the problem with keeping the accounting systems in synchronization with the project management system.

SUMMARY

Costs reflect work. Work only gets recorded if you assign resources to tasks in the schedule. If you assign these resources, then analyzing work is equivalent to analyzing costs in the project management software. Accounting systems tend to lag in reporting project costs. Thus, if you find out from an accounting system that a project is over budget, it has probably been in this state for some time.

Here is a guideline. If an organization has a good project management process in place, then it is easier and preferable to analyze plans with the project management software as the base. If, on the other hand, this is not true, then you are forced to rely on cost reports through an accounting system. In such a case, you must work backward from the costs into the schedule manually, since there is no automated link.

11 Succeeding with Projects Great and Small and Programs

Size can be measured in several ways:

1. Scope of the project—the more broad the scope, the larger the project tends to be.
2. Number of organizations involved
3. Importance to management—the greater the importance to management, the more attention the project gets. With more attention may come more organizations, more complexity, and, hence, greater size
4. Extended life and time of the project—the longer the time horizon, the larger the project tends to be.

Many of the methods we have discussed apply to projects of all sizes. Projects at either end of the size spectrum have some interesting characteristics. From experience, the size of a project is often a major determinant in how the project is managed, the likelihood of success of the project, and the extent of problems encountered. Then we will turn to programs, which are another type of project.

SIZE AND RISK

We often associate greater risk with larger size. Although this is generally true, there are exceptions. In projects of any size we treat risk as we have discussed in other chapters. Why is risk often higher in larger projects? Large projects tend to consume more resources across more of the organization. This means that there is greater competition for the resources between the line organization activities and those of the project. In ancient Mesopotamia and Egypt, work on canals and pyramids was scheduled at times when people were not involved in planting crops or gathering food. A core of people was maintained to tend the project during lulls in activity.

But there is risk in smaller projects. A project can start small and then grow into a large project. If the project structure cannot change with growth, there is greater risk and a higher likelihood of failure. In the Roman Republic armies were collected from volunteers. This was done in an ad hoc manner and was not particularly organized. It almost led to disaster in the Punic Wars with Carthage. The Roman Empire remedied this by establishing what is now known as a regular army. The army became a career path. In recent times this is borne out when we have a small pilot social experiment that works in one community. When the government attempts to expand it on a statewide or nationwide basis, different results occur, and failure is more likely. Why does this occur? People don't consider what happens and what changes in the project when the project grows. The lesson here is that when we do a small project, we always have to allow the possibility for the project to grow in an organized and consistent manner.

SIZING A PROJECT

The following are some of the dimensions of size:

1. Time. The longer the elapsed time, the larger the project and the greater the cost and resources.
2. Resources. This is the extent of resources required.
3. Types of resources. This can reflect the different organizations involved; it can also reflect the different functions involved in the project.
4. Underlying impact on the organization. This is the impact of the project on the infrastructure of the organization.
5. Use of new tools, methods, and technology. The more new things are used, the more unknowns there are.

As an example, let's examine the use of new methods, tools, and technology in several examples. Rome invented the use of cement. It was easy to work with and became widely used. But people were suspicious. Was it strong enough? Would it sustain buildings during earthquakes? For this reason, we find in many structures the use of bricks and stone as well as cement. The projects were larger. In some walls, the cement wall formed the core of the wall and an outer wall was built in more traditional ways. The cost and effort were then greater. As an aside, this is one of the main reasons why many Roman structures, roads, aqueducts, bridges, and walls remain today and are still in use. They were substantially overbuilt.

To avoid a massive project and its impact on organization and outlook, governments will often maintain the status quo. In nineteenth century England, the British Navy resisted moving to steam and coal from sail. The line commanders fought changes and for years used the methods employed in the Napoleonic Wars.

Today, although we tend to embrace new technology more rapidly, organizations still resist. Look at what happened to IBM with the rapid growth of microcomputers and downsizing from mainframe computers. The new technology can challenge existing beliefs. IBM continued to pursue traditional lines of products and proprietary systems at a time when other firms such as Hewlett-Packard were moving toward open UNIX-based systems.

CHARACTERISTICS OF LARGE PROJECTS

Table 11.1 gives a list of characteristics of large projects. Most of these relate to complexity. In some cases, such as contacts and getting status,

TABLE 11.1   Examples of Characteristics of Large Projects

The following list pertains to many larger projects. Which points pertain to a specific large project depend on the project.
   More subprojects and team leaders
   Project administration infrastructure
   Formal reporting process
   Greater use of formal software tools
   Tightly controlled and coordinated project plans
   More formal interface meetings between groups
   Full-time project managers
   Multiple locations involved in the project
   Multiple organizations involved in the project
   More complex tasks and work to be done
   More complex and difficult-to-use tools in the project
   Increased integration effort between work results of subprojects

there is more effort because more people have to be contacted. This is a linear increase. If you double the size of the project team, you essentially double the contact effort. But there is a multiplier effect. Many issues have more ramifications and nuances.

Larger projects tend to require a full-time project manager. In addition, there is much more administrative work in keeping track of work, handling requests for resources by the project team, and other work. This led to the establishment of the project office. The project office is the administrative arm of the project that does many of the things we have mentioned in earlier chapters. This is not new. Ancient Egypt had a project office with the builder and architects of the pyramids. In some armies, the military general staff can fill this role. Today, project offices can be found in large aerospace, government, construction, and banking projects.

Large projects tend to establish a project hierarchy. There can be a project committee that is composed of project managers of individual parts of the project. The overall project manager cannot easily contact members of a project team that are remote. This presents control problems and tends to encourage a formal project management approach. In aerospace, a large project is called a program. Within the program there are many projects.

Large projects tend to be more conservative and resistant to change. For example, the federal government has many commissions that have outlived their original purposes. Why do large projects seem to live a long time or seemingly forever? Complexity extends life; sheer size extends life. Another reason is that the project team somehow keeps it going.

Some large projects waste resources. With a large number of people involved, it is almost inevitable that there will be waste. The goal is to control it and keep it to a minimum. Here are some examples of waste:

- The large project team or part of the project team is put on hold while management makes some decision.
- Two different project teams are working on separate parts of the project. They do not communicate enough so their end products do not interface.
- Any large project tends to make a substantial set of assumptions about the organization, politics, technology, and marketplace, and other factors. The longer the project goes on, the more likely it is that the assumptions will be violated.

Large projects seem to have a greater tendency to fail.

- The project life is so long that the original need for the project disappears and the project fails.
- The project encounters a technical hurdle and after struggling fails; the original Panama and Suez Canals are examples of this.

These comments indicate that we should be reluctant to create large projects.

WHEN LARGE PROJECTS CANNOT BE AVOIDED

But a large project may be necessary and unavoidable. There are several approaches.

Normal Project

A large project can be treated as just another project. Some government agencies use this approach with mixed success. The difficulty here is that a large project requires a different type of management.

Superproject with Subprojects

This approach, which is more widely used, is to create a superproject. The superproject is an umbrella for a number of projects. There are several types of subprojects within a superproject:

1. Parallel projects. These are subprojects with limited dependencies and relationships.
2. Control project. This is the project office that attempts to control and manage the group of projects.
3. Infrastructure projects. These are projects that provide logistics support and may support tools and methods.
4. Integration projects. These are projects that begin after the superproject has started. Their focus is to integrate and combine results from several subprojects. We know from historical records that this approach was probably used in the construction of the pyramids. Stone and rock was quarried and then worked and then transported to the site. Rocks were then placed and integrated into the pyramid. The workers had to be fed and supported. Logistics support was enormous.

DESIGNING A LARGE PROJECT

Coming up with a structure for a large project is a challenge. Here are some suggestions for defining the project.

- We begin with defining the objective, scope, and environment of the project as we did in an earlier chapter.
- Instead of plunging into detail, we must carefully delineate the organizations that will be involved in the project as well as the areas of major risk.

- Think about the actual work itself. How should the work be structured? What things have to be done before others?

In terms of project structure, define a subproject for control. This is the project office. Identify any projects that will provide support in terms of tools, methods, technology, logistics, and so forth. Break up the project into pieces where basic work (not integrating work) can be done. Each piece should have one organization accountable for the work. What do you hope to accomplish in the integration of project pieces? You need to verify that the delivered end products from individual subprojects meet their objectives. This can be done by inspection, testing, and review. You will need to have a method for integrating the separate parts. The method may have to be supported by additional tools. The integrated pieces of the puzzle will have to be tested.

We actively plan how communications and networks will be employed. This includes allocating the plan tasks among the team members. An effective approach is to publish guidelines as a web page. Joint development with the team builds consensus. We have found that network use can help make a large project less formal and rigid.

SMALL PROJECTS

What characterizes a small project? They may only involve one or a few people. They occur within the standard line organization. They have a very limited duration and have a very small budget. There may be little risk. But small projects can lead to larger projects. If a small project is successful, it might be the model for doing many other small projects. An example is quality management. If it works in one department on a pilot basis, it may be extended to other departments. A small project may be progressing nicely. Management decides that this small project is a good vehicle to accomplish other goals. The project scope is then rapidly expanded. The small project may have end products that are required and are key needs of other, larger projects. Major projects of construction and engineering can depend on several small, critical projects.

TAKING A SMALL PROJECT SERIOUSLY

Small projects are started because there are specific needs—just the same as larger projects. Thus, there is still a need for project management. It may be less rigid and more informal, but it is still project management. Without structure and management, it can fail just as larger projects can. Small projects are very valuable to organizations for several reasons:

1. Small projects tend to be a useful training ground for future project managers.
2. Breaking up larger projects into smaller ones helps to increase control and accountability within the projects.
3. When a company wants to exploit some new technology, tool, or product, a small project is appropriate.
4. It is often possible to manage a small project within the line organization without establishing a separate project organization.

Labeling a small project as a project is important for several reasons:

1. It provides structure and some degree of visibility for the work to the organization.
2. It allows the organization to appoint a project manager so that accountability is present.
3. It gives attention to the project, and the project manager may be helped in terms of resources. After all, if no one knows about a project, it is more difficult to justify getting people to work on it. Disadvantages are potential overhead and too much visibility.

LEVELS OF PROJECTS

To address size and complexity, we can define four levels of projects, ranging from level 1 informal to level 4, a large, formal project. Levels provide consistency across projects as well as accommodate changes in size and scope. If we do not use levels or categories, we diminish our ability to manage projects in ways suitable to the project's characteristics.

LEVEL 1 PROJECTS—SIMPLE AND INFORMAL

A level 1 project is a very simple, informal small project. Duration is short; only one organization or a few organizations are involved. Generally, there are no pressing, important projects that depend on it. There is little risk and the organization would not suffer substantially if the project were delayed. To be a project in level 1, a project should meet most or all of these criteria. Size, duration, and most importantly risk and issues are determining factors in placing a project. Level 1 projects have a project manager and a very small project team (maybe, just the manager). What is needed is an organized method for handling issues, making project presentations, and keeping track of project data. A task plan is useful, but it is often very small and limited.

LEVEL 2 PROJECTS—SMALLER, STANDARD PROJECTS

In level 2 there is more complexity. More people are involved. As an example, we might have up to five people in the project. Project duration is typically less than one year. There is a single set of deliverable items. The number of organizations is limited. In general, this category can be characterized as larger than level 1, but lacks major issues and concerns.

LEVEL 3 PROJECTS—SUBSTANTIAL COMPLEX PROJECTS WITH RISK

Level 3 projects tend to be larger than level 2 in terms of resources and duration. But their main differences are complexity and risk. A large project that has limited risk could still be handled in level 2. Level 3 projects may involve more interdependencies between projects. They may also include more organizations. A level 3 project could involve technical risk and uncertainty. We set up a formal approach for tracking issues and for dealing with interfaces between parts of the project. We may have tasks established for specific interface issues. Level 3 projects are not so complex that they require a formal subproject approach as in the superproject. A project moves from level 2 to level 3 due to risk. It moves from level 3 to level 4 when it is clear that the single project approach is not working or when the complexity and risk keep growing.

LEVEL 4 PROJECTS—LARGE SUPERPROJECTS

This category is reserved for the large projects in which we must create subprojects and perform the management tasks discussed earlier in this chapter. Formal and automated controls are needed. Dividing the project into subprojects is part of the work needed. A project that grows in scope can arrive from level 3. It is usually evident that the existing project structure is straining under the pressure. A project can also arrive at this level because of problems, lack of performance, and other issues.

As a project winds down and subprojects are completed and phased out, then the project size shrinks. It is often appropriate to move down from one level to another.

HOW TO MANAGE THE LEVELS

It is not enough to informally recognize that different levels of projects exist. We need to define guidelines for placing projects in levels and for reconsidering when their level should change.

EXAMPLE

Hewlett-Packard is a manufacturer of technology-related products in medicine, computer systems, and other fields. The company, founded by William Hewlett and David Packard, was formed in an automobile garage.

The company had a decentralized focus based on specific product lines. When a division grew to over 1500 employees, it was split. As an example, the digital voltmeter division spun off the calculator division; the calculator division spun off a workstation division. Over time, the company grew and started to become centralized. It took longer to make decisions and the company lost its competitive position. By 1990 there was a crisis in the company. The founders returned to take on the turnaround project. This had to be done quickly due to both financial pressure and the fact that the founders were nearly 80 years old.

Let's consider this project in terms of levels. If they were to sit in a room and define changes themselves, then this would be a level 1 project. Involving a few advisors would make a level 2. Taking a centralized approach would move it to level 3. But the approach they took was to meet with small groups of employees across the company. They decentralized the company back out to the divisions. This approach has been very successful.

This example shows that one can place projects in different levels, depending on the frame of reference. Placing a project in a specific level does not just depend on risk or size, but also on the approach taken within the project. A decentralized project will have more subprojects and entail more coordination.

PROGRAMS: DIFFERENT MEANINGS

Over the years, different uses of the word *program* have been used in the context of project management. In the large aerospace projects of the 1960s and 1970s a program was a part of a project. We define a program as a series of repeating projects of the same type, or a project of indefinite duration. Examples of programs are

- Tax return preparation and filing
- An annual assessment of technology
- The annual budget cycle
- Quarterly progress reviews
- Continuous downsizing of organizations
- Performance appraisals for employees
- Carrying out a vaccination program
- Performing quality assurance on a range of projects

There is another type of program. It is the recycled project. This occurs when a series of similar projects is undertaken in sequence. Each project in the program is similar to the others. Examples of recycled projects are the development of consumer products, implementation of the same computer system in multiple departments, and installation of Electronic Data Interchange (EDI) and electronic commerce for a variety of suppliers.

In the area of consumer products, we differentiate program management from product management. Product management focuses on the life cycle of a product such as household cleaners. Program management is the standard development of new variations of household cleaning products.

The ancient Roman government developed a standard work breakdown structure for implementing Roman rule in newly conquered provinces. This included stabilization, setting up garrisons, building roads, constructing government facilities, and providing for water. This makes obvious sense. Otherwise, how could the Romans have colonized and pacified large parts of Europe and Africa in the relatively short amount of time that was required?

When we approach a series of projects as separate and distinct projects, we never learn from the mistakes of the previous projects. Also, because we know at the start that we are going to be doing the same tasks at different times, it is wise to establish an overall strategy, regardless of who will be assigned to the project.

What are some characteristics of programs?

- Although the same steps are taken at different times, the detailed information, people, and organization may change.
- Management has increased performance expectations because they are aware that it has been done before.
- Information from previous projects is frequently used in successive projects. Examples are long-range planning and competitive assessment efforts.
- The staffing that is allocated to a program is very small in most cases. This applies to planning and other efforts.
- The deliverable items or milestones for a program are typically very well defined.

HOW CAN YOU MAKE A DIFFERENCE IN A PROGRAM?

This is one of the "bottom line" questions. Why not follow the path of least resistance and just do the minimum amount of work to produce the deliverable items? Because you want to make a difference and produce a good work product.

Often you may have just one resource—you. At the meeting where you are given the assignment, ask several questions. What would they like to see in improvements this time? What were problems or issues last time? Did the milestone report have to be revised? This shows a need for improvement. By the way, management will not expect to hear much from you unless you have a problem. It has been done before so why should there be any issues?

What do you do next? Get an understanding of three previous programs or the last three times the program was undertaken. These should include the following:

- Reports and presentations from the work
- Project files containing working papers, memos, notes, and so forth
- Software or other tools used in the program

Organize the materials for each project in the same order. What is missing? Is it critical in terms of missing data? When you manage a program, you realize very early that with the tight schedule, you will need to start gathering data and organizing the final report or milestone as soon as possible. Get to work while it is fresh in your mind.

After reviewing the material, try to contact the person who was the previous program manager. Don't ask them about any problems or what to do. First, tell them what information and documents you have. Are they complete and up-to-date? Second, ask them for any advice on the process. What did they learn and what would they have changed? Underneath the conversation, try to get the answer to the question. Why are they doing something else now?

THE SCOPE OF THE PROGRAM

There are several areas of the program that should be given attention quickly:

1. Gathering the information. What are the sources for the information you need? Should you start immediately gathering data? From whom? When will it be ready for you? Are there any problems?
2. The method and tools. What should the method cover? Data gathering, organization of the information, analysis of the information, defining the audience for the milestone, and producing the milestone itself. Some sources and tools are external databases, software programs, reports from various departments, statements by officers of the company, audit reports, and so forth.
3. The end product or deliverable item itself
4. The use of Web pages to disseminate program information on a regular basis.

THE PROJECT PLAN

With the scope defined, you can start building a small project plan and task list. You should start with defining intermediate milestones. This will give you focus. Some of these might be

- Identification of data sources
- Completion of the data collection
- Initial analysis of information for completeness, correctness, and consistency

- Completion of initial analysis for the final milestone
- Completion of the first draft of the final milestone
- Completion of any presentation materials

Then start estimating the timing of the milestones by subtracting from the due date of the final milestone. Remember to allow for review time by management. This is often overlooked or assumed to be too short. Allow at least a week or two. Assume that you will have to develop a minimum of three drafts of the final milestone before the program is completed.

TAKE A BREAK: DEVELOP A QUALITY STRATEGY

A quality strategy is a definition of how the final report or milestone can be improved over several iterations. Note that you will not be able to make these improvements the first time because you are driven by the schedule. You are not given time to leisurely develop a quality end product.

What are some possibilities to consider in the quality strategy?

1. Expand the data collection to more sources, including additional external sources. People typically overlook external sources and databases.
2. Improve the quality of the information by finding better sources of the data. Get closer to the source or generator of the data. For example, instead of a magazine article, go to a company report.
3. Increase the depth and breadth of the analysis. Consider trend analysis over time.
4. Enhance the quality and understandability of the presentation of the results. Develop an outline for a presentation that will address management issues.
5. Adjust and expand the audience of the final report. Think about who in the organization might benefit from the report in addition to management (e.g., specific line organizations).

There is a trade-off between what additional quality we want and the cost and effort to review the work. Imposing better quality without thinking through the consequences of rework and measurement of the end product is a recipe for losing your credibility.

In terms of analysis, you could consider additional periods of time. You can consider more detailed information. The mathematical and statistical tools might be improved. To get presenation ideas, look at other similar reports. In terms of audience, ask yourself if there are other people who might gain insight from the report. Is it appropriate to prepare a short summary for a wider audience?

To develop some additional ideas for the quality strategy, you might review the previous report and ask the following questions:

1. What message does the report convey?
2. Is the message clear and supported by the analysis and information?
3. What is the reader or audience prepared to do after reading the analysis and data? Are there any missing steps? Could they come to widely different conclusions from the data? If so, then the analysis is incomplete.
4. Is the format of the report easily understood?

How do you present improvements to the process in the program? At the final report and presentation take a few minutes and talk about what improvements can be made. You should get a positive reaction. Not many people expected you to think of this. You might even give an example to show what can be done.

COLLECTING DATA EXTERNALLY

Internal data collection is similar to that for regular projects. Many programs require external information related to industry practice, financial data, and competitive data. Some of the external sources are industry associations, surveys, annual reports, form 10K financial reports, and so forth. You have several challenges here. First, you have to find the data. This can be a time-consuming process if you have never done it before. You should ask the person who collected data previously for sources. Also, you might consult a reference center or library.

A second problem is dealing with the information after it has been obtained. Information may not be consistent in detail, accuracy, or timing among sources. Because you cannot re-create exactly accurate and compatible data, we suggest that you make a copy of the document or the text that describes the data you are retrieving. In that way, you know how the data was collected and the assumptions on which it was based.

What are some common problems with external data?

- The date and frequency of the information are different. As an example, companies have different ending dates for their fiscal years.
- Amounts that are cited and that you are using may either be unadjusted or adjusted. Examples are currency adjustments, conversions, inflation, and physical units.
- Even the words used could mean something different to different people.

How do you handle these problems? One way is to read the entire source. Another is to contact the source and attempt to obtain more information.

ORGANIZING AND CHECKING THE INFORMATION

First, let's talk about getting ready to receive information. You should consider the manual and automated tools you will be using. This includes filing and logging in the information as well as automated tools. When you receive information, look it over. Here are some specific steps:

1. Compare it to similar information from the previous project. Does it make sense? Are the same headings used?

2. Did the information come in at the same time of the month or year as the previous time? Is it early or late? If so, find out why.

3. Read through the information to see what you can learn from this information alone. Next, if you already gathered other information, review this in light of the new information. Do this as soon as the information is received. Why? For consistency. If there are problems with information, you want to find them as soon as possible.

ANALYSIS TOOLS: SOME SPECIFIC SUGGESTIONS

We strongly recommend that for many programs involving data collection that you use microcomputer software. There are many different software packages. We are not going to comment on the specific software brand or make, but we can offer the following suggestions:

1. Identify your analysis software tools early. You may need to use statistical software, spreadsheets, graphics, and database software. We address graphics below. It is most useful if you can enter the data into a standardized database management system with the ability to export the data into a spreadsheet, graphics, or statistical package. If you enter data into a spreadsheet and you later want to do something else with it, you may have to reenter the data into the second software tool, which increases chances for more data entry errors as well as creating additional work.

2. Design and set up the software to receive the data in advance. When you are doing the design, think about both the analysis and the output. If you are going to create and establish formulas and so forth, use test data beforehand to verify that these work. You do not want to uncover computational errors using large amounts of real data.

3. Simulate the entire process with data from the last year or project. This will verify that graphs and reports can be produced.

4. Don't be afraid to use new tools that differ from those employed before. Here are two specific suggestions: First, the tools of the past may have been chosen because they were the only ones that the person knew. Second, new tools and versions of tools emerge annually. Newer software has more features and capabilities.

5. If you are going to use spreadsheets, make sure that any formulas you develop are checked. Lay out the format of the spreadsheet carefully. If you are using a database management system, then you will need to design and set up a database. Make sure that you put in edits for validity checking (e.g., range checks).

6. As you enter the information, make sure to check it manually. You will learn more about the information if you put "your hands on the data."

7. Make a backup of all data. Save data from previous periods separately as well.

8. You do not need to be an expert in the tool, but you should have more than basic knowledge. In the books, manuals, and sample data, look for examples that are close to your application.

DOING THE ANALYSIS

Now after selecting the tool and gathering some of the data, you are ready for analysis. Again, do not wait until all of the data is collected. Go all of the way through the analysis and produce final graphs and tables.

Some analysis guidelines are as follows:

1. Select standard, accepted methods of analysis. You will have to explain what methods you used. You want to use accepted methods so that no one will question them.

2. Verify that the analysis intuitively makes sense. For example, if some expenditure level is rising in the industry at 5% and your data indicates 15%, then there is possibly a problem.

3. As you perform the analysis, start writing down findings on a separate piece of paper or on the word processor. Set up a separate list for hypotheses. Hypotheses are your conjectures about why the analysis is coming out the way it is and what the data means. Even hypotheses that cannot be verified may be interesting later.

EXAMPLE: DOING A COMPETITIVE ASSESSMENT

In a competitive assessment you are collecting information on the activities and results of other firms that have something in common with your organization. They do not have to be in the same industry. They only have to share the same processes, issues, technology, or goals. Why do we do a competitive assessment? In order to improve our own business processes and the way we do our work. We also want to know how we are doing relative to the industry and to other competitors.

This leads to the first issue. Narrowly, we would interpret the assessment to be how we compare. On a wider scale, we would ask what it means and how we can improve ourselves with this knowledge. Competitive assessments are becoming more important in the 1990s as we move more into global competition.

After the issue of objective, the next one is scope. We could just consider firms in our own industry. This is a narrow interpretation. A wider interpretation is to consider firms in similar industries, firms overseas, firms who share the same technology, and so forth. We prefer a broader view. There are many sources of information. Don't just stick to annual reports of companies. There are many technical and management journals and magazines. Consult these. Look for articles written by employees of these firms. Collect qualitative as well as quantitative information. Find out what they attribute their success to.

The analysis tool should be a database management system because it can easily handle both textual and numerical data. Such a tool has the flexibility discussed earlier. A second analysis tool is graphics.

In terms of presentation you should define a structure for the report that includes what is accepted as classical numerical comparison. You should also include more advanced analysis and thinking that incorporates the wider scope of sources.

You should consider dividing your presentation into two sections. One section should focus on the classical analysis that they are expecting. The second addresses the new information. In your recommendations you should concentrate on issues such as the following:

- Which firms and organizations are worthy of further analysis?
- What business strategies seem to be working in other industries and how do they apply to you?
- What trends have you detected in other industries that apply to yours?
- What are some things that your organization should avoid?
- What are the key technologies likely to loom large in your industry?

The basic premise is that because you are making the effort to collect this information, you can get a lot more out of it by an incremental additional effort. You should measure in your mind the potential value versus the extra effort.

EXERCISES

We have defined four levels of projects. Take a series of projects about which you have read in the papers or at work and attempt to define the differences among the projects. Having defined the project differences, see if you can place each of the projects within a category or level.

Identify programs at work. Then consider how these programs are organized. Is there any effort to a structural approach? Is there any effort to have successive projects build on their predecessors? Is there any cross-fertilization of ideas between programs? If you answer no to these questions, then you could be in trouble.

EXAMPLE

Earlier we indicated that the overall project was very large. It included both reengineering and the establishment of a new project management process. Initially, some members of upper management treated the project as a level 4 project. This would have been the kiss of death to the project. It would have given too much management attention, created too many enemies too early, raised expectations, and created an administrative infrastructure that would have swamped an ocean liner. Hence, a substantial effort was made to separate the vision of what was being attempted with the tactical project. This was reinforced by the project management who reported separately on the big picture or vision versus the tactical project.

A challenge during the project was to maintain the vision while at the same time keeping an eye on the prize—tactical reengineering and project management. The vision was enunciated to the levels below upper management very few times. When asked, the project team members would say that the project overall was to do the same steps in different organizations. The focus was on linear change, rather than exponential change, which would have been far more threatening.

In the context of this chapter, the overall vision would be level 4. The project management part of the work was generally level 2; the reengineering effort was level 3.

SUMMARY

Projects and programs share many of the same characteristics and methods. Differences often relate to addressing issues and changes within the individual projects and programs. It is important to be aware of political realities as to how to cast a project properly in the levels. The benefits of having project levels are to provide flexibility, support a lower profile political approach, and to indicate structure to management.

12 Product Management

A product is an item such as a car, a consumer item, a service, or something that people can use over time. Products have to be designed, developed, marketed, managed, enhanced, and eventually replaced by some new product. In a book on project management, we might wonder why product management appears here. First, product and project management have much in common. Second, the development of a product requires a project effort. We could have both product and project managers.

Unlike a project, which eventually concludes, a product continues after it is developed. This is a major difference between products and projects. This chapter explores product management from a project point of view, which should shed more light on some aspects of project management. We are not attempting to explore all of the details of product management; rather, we will examine some aspects of project management contained within product management.

THOMAS EDISON

Thomas Edison is viewed as the individual responsible for applying the generation of electricity to consumer and commercial applications and products. The interpretation is often that he was a theoretical inventor. Although this is somewhat true, it does not reflect where he devoted most of his life. When electricity was demonstrated and explained, many people did not think much of it. After all, what could it do? People had candles and gaslight. You rose at dawn; you worked; you went to bed at or shortly after dark. What could electricity do for you? It appeared to be dangerous.

Edison spent time developing electrical applications that today we call electrical appliances. Beyond lighting, he considered washing machines and a whole range of other applications. In fact, records show that he and his staff spent much more time on applications of electricity than on basic research. This may have been spurred on by newspapers that questioned the benefit of electricity with few applications. After all, people had mechanical devices and gas lighting, so why did they need electricity?

Edison was a project manager for these applications as well as for basic electrical engineering. More importantly, he was a product manager, because so much of his life was dedicated to electrical applications.

CHARACTERISTICS OF A PRODUCT

A product has a life cycle. It is first thought of as an idea. Figure 12.1 gives a chart of the stages of life of a product. After the idea is conceived, the hard work of designing the product begins. The product must be supported by strategies for manufacturing, marketing, distribution, and management. A product tends to have a much longer life than the initial development project. In addition, the product life cycle goes through successive project life cycles where new, enhanced versions of the product are developed and sold. In that way, a product life cycle is actually composed of a series of projects. This interpretation only makes sense if we consider the distribution, marketing, and support to be projects. Otherwise, we would simply have a series of nonproject activities sandwiched in between projects.

The development stages of a product in terms of design and construction are very similar to that of a project, as Figure 12.2 shows. But there are unique differences:

- The product will just be starting its life cycle in marketing and distribution when the development project ends.
- The product may be one member of a family of similar products. Examples are different flavors of the same ice cream. Thus, we will consider product families.
- The person who manages the product will tend to remain with the product during most of its life cycle. A project manager may see the

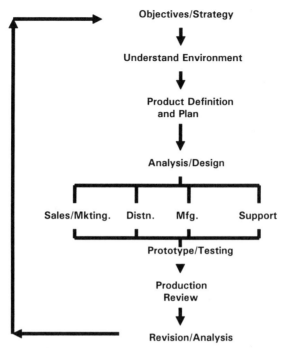

FIGURE 12.1 Example of stages of product.

product through development, but will not be involved in supporting it later. That is why the product and project managers for many products are different people.

There are also similarities. First, the product manager faces some of the same issues and conflicts that are faced in an everyday project. Second, the review and control process for product development and management have many things in common with project management. Third, both product and project management can use some of the same methods and tools.

A product is often the end result of a project. Whether the product works or meets its original requirements depends on more than just the original development project. It depends on successfully managing the product after development. That is why we tend to view these successive activities as other related projects.

CONCURRENT ENGINEERING

Over the past few years there has been increasing concern over the sequential nature of developing products and in doing product manufactur-

ID	Name
1	1. Prod Objectives
2	1.1. Product Objectives
3	1.2. Competitive Assessmt.
4	1.3. Internal Assessmt.
5	1.4. Product Strategy
6	1.5. Document Objectives
7	1.6. Review Objectives
8	1.99 M: Product Go-Ahead
9	2. Detailed Product Plan
10	2.1. Product Definition
11	2.2. Product Plan
12	2.3. Project Plan-Development
13	2.4. Document/Review Defn/Plan
14	2.99 M: Product Go-Ahead
15	3. Product Design
16	3.1. Marketing
17	3.2. Sales
18	3.3. Manufacture/Assembly
19	3.4. Distribution
20	3.5. Product Mgmt.
21	3.6. Completion/Review of Design
22	3.99 M: Product Go-Ahead
23	4. Prototype/Testing
24	4.1. Prototype Construction
25	4.2. Market Test Planning
26	4.3. Conduct Market Test
27	4.4. Analyze Results
28	4.99 M: Product Go-Ahead
29	5. Setup Production
30	6. Setup Marketing/Sales
31	7. Setup Distribution
32	8. Define Mgmt. Reporting
33	9. Product Roll-Out
34	10. Production Review

FIGURE 12.2 Sample work breakdown structure.

ing. One concern is that the product design is compromised and changed in the manufacturing process. Another concern is that the product design as built does not meet the needs of the market. The field of concurrent engineering has arisen as a result of these concerns. Concurrent engineering is the early involvement in design by customers, suppliers, marketing, sales, purchasing, manufacturing, distribution, and other groups in the company.

The purpose of concurrent engineering is to obtain a better product that meets market needs at a lower cost. A number of firms have embraced concurrent engineering with great success. An example is the engine development for the Boeing 777 aircraft. Another example is the Chrysler LH sedan. There are other successful examples ranging from medical products to computer keypads to moldings. Product development cycles that took over five years have been reduced by as much as 50%.

Project management is a critical method in successfully carrying out concurrent engineering. The entities that are involved in the design have to be coordinated and managed. In addition, methods and tools have to be managed. Project management has been cited as a critical success factor for product development using concurrent engineering.

EVOLUTION OF A PRODUCT

A successful product typically evolves over time. New technology and the experience of earlier versions provide fodder for new ideas. For example, the original Sony Walkman® started as a simple radio. Later versions evolved into being resistant to sun and sand, playing compact disks, and even supporting television. This is product evolution. The product evolves.

People may learn something new during the development process and find ways to improve the product prior to introduction. This is true with new airplanes that make their journey through many different versions over a period of up to several decades. For example, a military airplane such as the F-16 will stay in the military inventory for over two decades. On the outside, it looks the same, but inside the avionics, hardware, and software are very different.

We see this in archaeology where arches, castles, turrets, bridges, and other structures evolved in both design and construction. Think of these as the evolution of a specific product. In ancient Rome the skills and techniques of building and supporting aqueducts improved during the early years of the empire.

What does evolution mean for project and product managers? For the product manager, evolution means that the product manager must devise an evolution strategy and plan how the product will evolve in an organized manner. For the project manager, there must be flexibility established and maintained in the design and development of the product so that later versions of the product can be accommodated. Viewed in this way, developing a product is a specific category or type of project to a project manager. The development of the initial product or a later version is just one of a series of development cycles that have to be endured or addressed.

PRODUCT FAMILIES

A product family is a set of related products developed by a single organization. The products are related through performing the same function (as in flavor differences), being successors to each other over time (as in evolution), or having manufacturing, marketing, or organization characteristics in common.

When we consider product families, we often end up with more complex projects because a single project may include several different products. Consider, for example, a marketing project for several different varieties of shaving cream. The project must address the different varieties and the different competitors as well as differences in preference among market segments.

GETTING A PRODUCT ORIENTATION AND WHAT THAT MEANS

In addition to consumer products and everyday items, we can take the objectives and results of projects and sometimes treat them as products. For example, suppose you are a software manufacturer or publisher and are working on a spreadsheet program. The project is to develop the software. However, you know that you will eventually want versions for different computers and operating environments. This is a product family. Since there will be multiple versions of a product, you should be careful in managing the development project. There are several reasons for this. It will save time and effort if you can use the same methods and tools for variations of products in addition to the basic product. If you don't build the product well, it will be difficult to implement new versions and variations later.

Taking a product orientation toward our work has value in that we will plan, work, and manage more carefully than if we thought of it as a one-time effort. A product orientation tends to support quality and control more.

PRODUCT MANAGEMENT

Product management is similar to project management with some additions. A product manager has to not only worry about the particular state of the product, but also about the next stages of the product. Product managers have to keep up with other, competitive products. If there are previous versions of the product, these have to be tracked. The scope of the projects that a product manager is concerned with tends to widen. Some say that the product manager is more like the manager of a superproject.

A product manager typically is not the project manager for the development of the project. The manager may be project manager for marketing and distribution, but seldom for development. There are several reasons for this. The development of the product is viewed differently than the perspective of the product and project manager. The product manager focuses on getting the product out and also on the long term. The project manager centers attention on the project and is not as concerned with what happens after that, which is a more narrow view.

PRODUCT VERSUS PROJECT PLANNING

A product plan usually describes characteristics of the product. There is usually a marketing analysis in terms of the potential market for the item and functions and features. A competitive assessment might be included. The project plan, by way of contrasts, lays out the objectives and strategies, but it is internal. The justification for doing the project has sometimes been made before. Product plans often must convince management to support the development, deployment, and enhancement of the product.

The scope of product plans covers the product life cycle as well as development. In many organizations a project manager provides a project plan for development of the product to the product manager for inclusion in the product plan. A product plan may also identify very specific issues relative to details of the product. These things would be issues within the project plan and project management.

RELATIONSHIP BETWEEN THE PRODUCT MANAGER AND PROJECT MANAGER

During the development of the product there is a natural tension between the two managers. Suppose, as an example, a design issue comes up in the project. There are several alternatives: ignore it, implement it, or consider it more analytically. The project manager will want to have the issue resolved because it may impact the project and it also carries increased risk. On the other hand, the product manager may wish to maintain flexibility and leave options open until as late as possible. This is a trade-off based on the time horizon and management perspective.

Another example occurs if product development begins to lag. The product manager has distribution, marketing, sales, and other plans in the works and even in process. Delaying the product would have a substantial impact on these plans. The project manager wants to do the job right so the trade-offs must be balanced.

PRODUCT DEVELOPMENT PROJECT PLAN

A product development plan reflects the wider scope of product development. Figure 12.2 gives an example of a work breakdown structure. Note that market research, marketing, competitive analysis, packaging, support materials, and other items are included. Note too that there is an integration subproject that attempts to tie these together. Figure 12.3 gives a Program Evaluation and Review Technique (PERT) chart for the project.

For many products, the product manager acts as a project manager for some of the subprojects. A different project manager would be in charge of developing the product. In a team development approach, the specifications, design, and development are overseen by a management team. Each team member would act as a project manager for parts of the overall project.

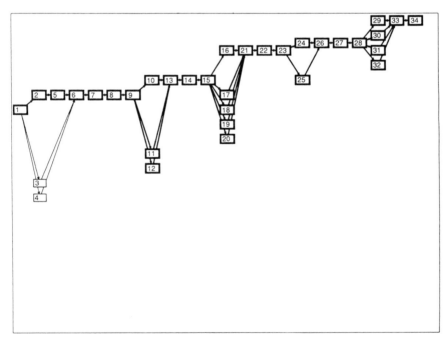

FIGURE 12.3   Sample PERT chart.

Each product is unique. This could be interpreted by someone into saying that the Work Breakdown Structure (WBS) and project templates don't apply. Don't buy into this! High level tasks and summary tasks can be standardized across many products of a similar type. Detailed tasks can be customized to reflect the unique product. The claim of uniqueness is sometimes used to avoid project management and control.

LIFE OF A PRODUCT MANAGER

The life of a product manager is often a study in hectic discord in that on many days the project manager's time is split among a variety of different types of activities. Let's examine some of the more commonly encountered activities.

New Ideas

The product development team has a new idea to enhance the product. The product manager must address several questions. How long will it take? How much will it cost? What parts of the design are impacted? What will the enhancement do to the features and capabilities of the product? How will the market place be affected? Should it be deferred until a later version of the product?

Problems

Product testing indicates that the product needs to be reworked prior to release. This could delay product launch or introduction by a month. You already have purchased advertising space in several magazines. Seemingly small events such as these can have a dramatic budget impact.

Competition

A competitor is coming out with a new version of their product that competes with yours at a lower price! You have a earlier version of the product on the market and a new version under development. What actions should you take? What about pricing? If their new version has features that yours in development lacks, should you do the enhancements and then suffer the delay? How will market share erode?

These are but three examples of how marketing relates to product management. These indicate that we need an overall strategy and plan that is up-to-date. Another lesson is the impact of a wrong decision. In many projects, deciding on an issue and then finding out you made the wrong decision can be reversed or adjusted. But in product management, the world is often not as kind. Once you decide on a marketing campaign, the die is cast.

As a product manager you also have to learn from experience and information collected. As you receive feedback on the product in testing or in use, you may choose to make changes in the product. You may want to correct problems. This is easier in production processes where adjustments are possible. An example is software publishing where a firm can release version 2.01 to fix problems in version 2.00. It is much more difficult in consumer products such as soap, soft drinks, or other items. That you can in fact learn and benefit from this is shown by Coca Cola's experience with Classic Coke and the new Coke. By staging a competition between both soft drinks, they increased market share. Thus, the new flavor mistake was more than corrected by the new marketing strategy.

PRODUCT LAUNCH

Launching a product means that the product has been developed and tested. The activities that build up, coincide, and immediately follow the introduction of the product are included in the product launch. In many organizations the product launch is a project in and of itself. It may need a separate project leader. This project requires substantial coordination skills over a very short time period. Timing tends to be very important.

We have already touched on some of the elements of launching a product. What are some of the problems that can be encountered?

1. A previously unknown error or problem with the product surfaces in the launch period. Do you delay the launch or do the launch and then recall or replace the items that were delivered?
2. Marketplace and competitive events occur during the launch period. How do you respond in the advertising and what you are doing in the launch?
3. Distributors and dealers for the product are complaining about pricing and discounts related to the product. Do you dare reprice the product? Do you give deeper discounts?

Even with the product finished and on the shelf, there can be problems and issues. And as you can see from the examples, the issues and questions are often intertwined. How a product is priced impacts the market segment you are seeking. Advertising and the distribution channel relate to market segment. Everything is tied together.

MEASURING YOUR PRODUCT

An issue that occurs after and during development is an assessment of how good or bad your product is. This is similar to testing in a project.

There are differences. Project measurement tends to be well defined and internal. Product measurement can be both internal and external. Product measurements can change over time depending on the marketplace and audience.

In project measurement, the end result may be some minor changes and adjustments. In product measurement there is a much wider range of possible decisions involving design through sales.

PRODUCT MATURITY

Let's call the end product of a project a system. The system that was constructed in the project is then operated and maintained after it is put into operation. Additional work may have to be done on the system. Usually, this is either a small set of tasks, or a limited enhancement project. There is no consideration of a follow-up on projects of the same size and scope.

Things are different in product management. We want to make money on the product, but we also want to phase out the product with a new version or replacement product. The strategy is one of continuous change and evolution. The product manager wants the system or product to be successful, but is more concerned about the entire product family over a period of years. Many products are issued and produced when we know that the goal is that they become obsolete relatively quickly. This occurs and becomes a pressure point as the market is saturated with the product.

DEATH OF A PRODUCT

A product can die a number of different deaths. Projects end when they are completed, are changed, or are stopped. Some of the reasons for the death of a product are as follows:

- Loss of demand for the product. People don't want it anymore. There is not enough demand to support the production of the product.
- Replacement of the product by a new version or product. The same manufacturer or competition effectively replaces the product in the marketplace?
- Relegation to a niche. The product is still produced, but it is done so for a very limited market. This means that the product is effectively removed from the marketplace. Examples are items that you want, but that are difficult to find.

What are the signs of death? Product sales, perhaps, have been declining. No one has been able to think up ideas for the product. It is mature in the sense that it has already progressed through several versions and now is in the last stages. This is like a car that has had several face-lifts and now is to be replaced.

A product manager faces several challenges when a product is in decline or is mature. A strategy must be devised for determining what should be done for the product. Some options are:

- Spend money and determine what could be done to extend the life of the product. The product could be discounted in price; other options can be added to the product to reduce the effective price.
- Invest in making changes to the product to increase its appeal. The problem here is that the "production line" or development team may have moved on to other work so that restarting the project is prohibitive.
- Look around for a product replacement. What new product could be acquired or offered that could be sold as the logical successor to the product?
- Milk the product until sales or demand disappears. This can be done with books when a publishing run is performed. Books are stored and sold out of storage. Decisions are made later on additional printings and editions.

Whatever the strategy, the product manager has to develop a plan to support the strategy. The plan has to be marketed to management. Design and implementation of the plan are similar to the development of the product that we discussed earlier.

PITFALLS IN PRODUCT MANAGEMENT

There are a number of potential problems and pitfalls that product managers should avoid. Many of these apply to project managers as well.

Organization and Structure

Do not assume that you can run a product (or project) in an informal way and succeed. If you do not have communications and structure, you will be more likely to run into more problems and issues. These issues will then take more time—the penalty for not being organized.

Strategy

Always keep the product strategy in mind. There will be many times when you will be tempted to sacrifice the strategy for very good short-term goals. This may in turn end up being your undoing as a product manager.

Long-Term View

Both product and project managers have to take a longer term view and not just focus on short-term issues. This means that not only must the strategy be kept in mind, but also the detailed plan for the next year.

Flexibility

Be flexible. In many products and projects there is intense pressure to freeze changes and prevent new ideas. People are trying so hard to perform that they cannot tolerate change. You should encourage new ideas. When you cannot figure out how to use a new idea that has merit now, then see how it can be fitted into the strategy.

YOU ARE A PRODUCT

You may not think of yourself as a product, but you are. You are in fact a product family. Over the years, you will learn new skills and gain additional knowledge and experience. You will perform different work using different technology over time. Most people do not recognize this and some are not even aware that it is happening, but it is.

Think back ten years. What new technologies have emerged? How has your life changed? What is your average day like today versus ten years ago? You are different now than before. More precisely, you are a different product. You are your own product manager. By that we mean that you have to develop your own skills, measure yourself, market yourself and your skills to others. You are the best one to do that. You can see the comparison in books. If you pick up a book on product marketing and management, look through the table of contents and think of yourself as the product.

Now we are not attempting to delve into psychology, motivation, or other areas, but we can point out the benefits of taking a systematic project management approach to yourself as a product. We can see that you need to have a product strategy. Where are you going over the next five to ten years? What new skills will you acquire? What new experiences can you add? Or, is it more of the same experiences? Are you a mature product becoming obsolete? Some people get complacent and think that because things are going along, they do not need to change. They do not learn new skills; they rest. We know that this is a mistake, but it is so tempting to sit back and enjoy life when you have worked so hard to get where you are.

If you take the product approach, you cannot sit back. You have to improve the product continuously. You have to have a strategy for improvement. In that regard, a human being is unique. Not many things, not even computers, are as flexible. A person can work in any number of fields in a given period of time. People can also cope with change more easily than other structures.

Change should be organized change. Change for change sake will often result in failure. In product management we seek a structured approach for change. The elements of project management that apply here are

- Develop your strategy.
- Develop a project plan to support the strategy.
- Measure your progress and yourself.
- Be prepared to deal with and address issues.

This last point is important. Many people attempt to go through life avoiding issues and unpleasantness. Yet these types of experiences allow us to grow and improve. Coping with problems and situations can give us immense satisfaction.

EXERCISES

Start with the previous section and think of yourself as a product. Your goal is to evaluate yourself now and compare it to the product you were in the past. What are the major differences in terms of experience, skills, knowledge, tools, techniques, and so forth? What types of projects, situations, and experiences contributed the most to your growth as an individual? What we find is that a few selected incidents made major impacts on our lives. This was especially true of Sir Winston Churchill and Richard Nixon. Both were failures and successes in a twenty-year period.

After evaluating yourself, start thinking about other products around you. What products have you used for some time that are now harder to find? What products have essentially remained unchanged? What products have disappeared? Make a list of each of these and put the list away. Each year get the list out and see what has changed. This is an interesting approach to visualize change.

Another approach to track evolution and change is to keep a diary. You would isolate a typical day each year and write down in detail what happens in your day. By comparing your notes between years, you can see what has changed.

EXAMPLE

We can think of the overall superproject as a product with several variations. Over time, the project team used this approach effectively in internal communications and in explaining why reengineering and project management approaches differed between different business divisions.

In parallel to the project or product, some business divisions attempted to circumvent the project by going out on their own to attempt reengineering and set up new project management processes. These were viewed by the project manager as competing products. This approach helped to diffuse emotions in meetings with upper management. The concept was that in many organizations there are different approaches. Each approach can be viewed as a product competing for shelf space in a supermarket. In that way, competition could be viewed as healthy and not counterproductive. Politically, because it took time to get to all divisions, it was inevitable that some tried to go out on their own. The project team added to their tasks active tracking of the competing products (just as in the competitive assessment mentioned earlier).

The products evolved over time, like the F-16 avionics systems mentioned in this chapter. Divisions addressed in earlier projects were retrofitted in terms of methods and received the corresponding benefits. These efforts were done at periodic intervals of about one year.

SUMMARY

This chapter has been a partial diversion from strict project management. It is useful because all products involve projects of some type and most projects end up producing systems or products that are maintained and supported for an extended period. It also serves to indicate how large projects with many parts or subprojects can be treated as products. Experience has shown that it is often useful to treat reengineering efforts in a product way so that the improved process is better defined and understood as a product.

Product management differs from project management because the life span of the product exceeds that of a project. In addition, product management tends to be focused on a wider set of factors than a project. External factors such as demand, the marketplace, and competition play a large role for a product, but may play a lesser role for projects. Product managers share some of the same issues and problems as project managers. Both require an organized approach. Both can use some of the same tools. However, the product manager often faces greater complexity in the interrelationships between tasks and issues.

13 Managing Issues in Projects: Problems and Opportunities

PROBLEM VERSUS OPPORTUNITY

Recall that a problem is a situation that threatens the project and has to be addressed by the project team and leader. We seek not only to resolve problems, but hopefully turn something negative into something positive. Opportunities represent means to improve the project or its results in some way. Opportunities may not have the same urgency as problems, but they are important. We have grouped both problems and opportunities into issues.

STEPS IN ADDRESSING ISSUES

We will use the following steps in defining and addressing issues.

STEP 1: IDENTIFY THE ISSUE

By identification, we mean that we have studied and understand the issue in terms of the following: (1) symptoms of issue; (2) underlying parts of issue; (3) potential impact of issue on project; (4) urgency of the issue; (5) priority of the issue. Note that urgency and priority are not the same. Urgency is the importance to the project. Priority is the order in which you will get to it as a project manager. If you are still collecting information, then an issue can have a high degree of urgency, but only a medium level of priority.

We differentiate from symptoms that are visible and the parts of the issue that may underlie the symptoms. This takes analysis because there may be multiple issues involved as well as multiple symptoms. This is a many-to-many mapping. Therefore, when we identify one issue, we may actually address several at one time. Our first suggestion here is to review all open issues when you look at a new issue to determine if it really is new or if it is a retread of a previous one.

STEP 2: GET CONSENSUS ON THE ISSUE

As work is going on to define the issue, it is important to share your ideas with others on the project team and with management. People may need to be informed about the issue and its ramification. Otherwise, if you propose an action or solution, you risk having the solution rejected because it solves a nonexistent problem or it does not fit other valid, but different interpretations of the situation. Consensus is important to gain support for action.

STEP 3: DEVELOP ALTERNATIVE ACTIONS FOR THE ISSUE

Once an issue is defined, we can start to define alternative actions. Many of us want to plunge right in and select an approach. This is a mistake. There are certain alternatives you should always consider for any issue.

1. Do nothing. Do not take any action. Let the situation stand and collect more data. Here you would identify the advantages of waiting.

2. Address the issue without any additional resources or changes to the schedule. This is addressing the issue within the confines of the project.

3. Address the issue with changes in the project, but with no additional resources. This may involve redirecting resources and changing schedules.

4. Stretch the schedule with the same resources. This approach is to provide the same resources for a longer time. The schedule will be modified.

5. Add resources, but keep the schedule. This is a typical response when there are tight deadlines and where additional staff can make a difference.

6. Add resources and stretch the schedule. This is obviously an extreme alternative.

After you have defined what these and any other alternatives mean to the project, you will need to relate these to the issue or issues at hand.

Then you can determine what impact and benefit will occur when action is taken on the issue. To provide structure for defining impacts, consider impacts in terms of the following list:

1. Additional risk in implementation or to the project work
2. Changes to the project structure
3. Changes in roles and responsibilities of members of the project team
4. Changes in interfaces with organizations
5. Alterations in methods and tools

Having defined the alternatives and eliminated some based on financial or schedule constraints or other factors, you need to see if these alternatives can be part of other changes that should be made in the project at the same time. See if you can define a strategy for changing the project composed of a number of actions. We discussed strategies earlier. A strategy here is the same except that it provides motivation for making changes.

At this point we have a series of alternatives that fit under strategies and that yield various impacts. You have filled in the diagram in Figure 13.1.

STEP 4: DECIDE ALTERNATIVE, SEEK CONSENSUS, AND DEVELOP AN IMPLEMENTATION PLAN

If you cannot decide on a specific alternative, then perhaps you should collect more information and leave the issue alone. The implementation plan includes how it will be undertaken.

STEP 5: MARKET AND SELL THE APPROACH TO MANAGEMENT AND CARRY OUT IMPLEMENTATION

Why go to all of this work? Because you have to prepare a case and approach even if it is to do nothing. You will have to make an argument for additional resources, extended schedules, and you will have to explain why there is no need for action.

Alternative	Resources	Schedule	Impact	Risk

FIGURE 13.1 Table for alternatives. Table entries are filled with text describing effect and so on. Impacts can be internal or external. Risk can be identified by type or severity.

USING NETWORK TECHNOLOGY

Analysis and data collection of issues should not be performed in a vacuum. With the network we can start a dialogue about the scope, symptoms, nature, alternatives, and potential actions on an issue or opportunity. To help get this started, you should establish a model or example on the network. This can be published as web pages for people to review. We have even used chat rooms to discuss issues related to technology, methods, and tools.

SYMPTOMS

Some examples of symptoms are listed in Table 13.1. When you observe a symptom, step back from the project and ask why this was not noticed before and whether there are additional symptoms. Do not jump in and attempt to define the issue immediately. We will talk more about this with project change. We will now consider some issues individually. Keep in mind that several symptoms can appear simultaneously.

You should assume responsibility for all of these symptoms of issues because you as the project manager should have been on your toes. Don't blame management, the project team, or the methods or tools. If you get to the issue early, then you can, on the other hand, take credit for getting it recognized and fixed.

COST OVERRUNS

Although the total budget has not been spent, the cost of the project so far exceeds what has been budgeted. Sounds like a direct situation. What

TABLE 13.1   Symptoms of Problems

Lack of discussion about project
No discussion of task status
Difficulty in getting resources for project
Difficulty and elapsed time in getting management review and approval of work and
    actions
Lack of clear milestones for project
Lack of interest in project outside of the project team
Project manager seems to have problems
Project team members are reassigned
Project team members are trying to leave project
Interfacing and dependent projects are attempting to become more independent
Contingency plans to the project surface
Project objective and scope get cut back
Management vacillates on need for information on project.

is the cause? Where is the money going? If labor on the project is fixed and is not the source, then there may be unanticipated expenses that are accruing to the project. If people are working overtime on the project, then personnel time may be an issue. If personnel time is an issue, then you need to determine if people are spending too much time on assigned tasks or spending time on tasks that are not in the project plan. The latter situation occurs in some projects where the project manager has created a summary plan with a lack of detail.

Although we might suggest that you look for ways to reduce the budget, this will only help address symptoms. The causes of the problem in terms of poor planning, missing tasks, underestimates of work, and so forth are still going on. We would suggest that before any cuts are made, the project plan should be reviewed and updated in terms of estimated effort. The task list should first be reviewed. After this is shown to be complete, then you can review the detailed schedule for each task.

LATE SCHEDULE

You are still on budget, but the schedule is late. You may observe this when you are doing an update to the plan. It may come to you when you have seen that three tasks will be late. What do you do? First, analyze the plan to determine which tasks are late and what characteristics they share. Do they involve the same people? Are they based on using the same method or tool? Are they interdependent? Look for a pattern.

If there is a pattern, then you can focus attention on the subset of tasks that are part of the pattern. Look at the tasks for completeness and detail. Try to get at why there is slippage. Beyond changing the schedule, should the organization of the work be changed? We could ask other questions related to the roles and effectiveness of people. If after this analysis, you can focus on specific tasks, then you might want to pay greater attention to the staff who are working on these tasks. Not just observing, but getting involved positively.

If there appears to be no pattern, look at the project overall. Is there a problem across the project in terms of organization, estimation of effort, assignment of people, and task structure and dependencies? Have a series of tasks been underestimated? You are trying to look for systematic problems in the plan as opposed to identifying individuals and tasks to monitor and work with more closely. Another clue is to identify rework and extra work that was unplanned.

Note that we have mentioned looking at the project several times. You need to do this to determine the various ingredients of the issue. You could be making a mountain out of a molehill by not looking at the project first. We have seen people make errors in entering project data into a computer system. Then when the new schedule came out, they ran around in a panic.

When they finally sat down and corrected the error, they had already done harm to the project.

SUBSTANDARD PERFORMANCE

Let's suppose that the milestone for a specific task or set of tasks is not of good quality or is not complete. There are two parts to this issue. First, you need to take action on the milestone. This means improving or reworking the product to make it acceptable or better. Second, you have to determine how this happened and why you were not alerted to it earlier. You should review your involvement with the staff who worked on the tasks that produced the milestone. You should also review the skills and techniques that the staff used during these tasks.

Thus, as with other symptoms, you are trying to do several things at one time. You should involve the project team in this effort. As opposed to asking how it could happen, divide your effort into the work itself and then into the process for managing and checking the work. You need to get the team members' agreement on what steps will be taken to monitor the work in the future.

RESISTANCE FROM DEPARTMENTS

At first glance, this appears to be an issue. But it is often a symptom for deeper issues in the project. How would you detect symptoms between the project and departments? A hint could be dropped by a member of the project team from that department. Another way for this to surface is in contact with a department. There are many possible causes for an intradepartmental issue, ranging from simple miscommunications to a major dispute concerning a method or resource.

Because the symptom is not within the project, there might be a tendency to treat this lightly. Don't make that mistake. You need to uncover and resolve the source of the symptom as soon as possible. The project manager acts as the liaison contact for the project. Also, many of these symptoms often can be resolved by handling it as a misunderstanding.

We suggest that you start with the project team member from that department and attempt to find out as much as you can. Then you should think about the communications between the department and the project. Before going to the department manager, you should go to your own manager and indicate what you are doing and how you are going to handle it. When you visit the department manager, take along an updated project plan as well as a clear idea of what their employee(s) are doing on the project. This discussion reinforces the benefit for communications with line managers whose employees are on your project team.

STAFF TURNOVER

Turnover and the rate of turnover are functions of the size, scope, and duration of a project. Some project we have seen have gone on for several years. At the end of the project there was not one person on the team who was on the project at its inception. We have discussed the need for having part-time and temporary team members who come into the project, perform a specific set of tasks, and then leave. Turnover here is not expected, but should be timely so that the project budget does not suffer.

This leaves a core set of people on the project team. These individuals provide the continuity and strength of the project fiber. If they are leaving the project unexpectedly, then there may be a problem. Why do people get dissatisfied with being on the project? Some reasons we have encountered are as follows:

- No apparent relation of project to their career path
- Conflict between project and line management
- Inability to communicate with rest of team or leader
- Ineffectiveness because of lack of knowledge and training
- Perceived lack of interesting work

Often, the situation could have been prevented if the project manager was in better communications with the team members. However, there are legitimate reasons for someone to leave a project:

- They are needed on another project.
- They do not see remaining on the project as a benefit to their career path.
- They may feel that they have done as much on the project as they can.

The project manager should be in close communication with the core project team—every day. The project manager needs to be sensitive to the person's problems and needs. A good project manager should spend a lot of time listening to people. Again, we stress the importance of communications. No project manager has a 100% success rate here. People will still leave. You cannot always get to everyone at times of substantial stress and when you are spending time working on other areas of the project. You cannot always visit their office. Of electronic and voice mail, we suggest that you use voice mail to call people and stay in touch. Your voice can express your care and concern better than a dry, typed message. Another thought is to use neat, handwritten notes.

Even with all of this, let's suppose that a project member is going to return to their department. What do you do? You first have to make a clean transition and not have a lot of loose ends. This is in the interest of the team member. They want to go onto new work and do not want to be called back to work on the project. Try to have the transition occur before they leave. This gives a more gradual phase-out than a sudden break in the middle of a task. However, there are still likely to be tasks that linger. You should get in the middle of the transition. Review the work and status with the departing member. Be there when a new person is assigned to the tasks and takes them over. You should also write a detailed letter or memorandum to the manager about the employee and his or her performance and contribution. Try to be very specific and direct in this letter. Copy the employee and the project team. They will see how someone is treated—fairly.

TEAM CONFLICTS

A project team is sometimes compared to a family. In any family there is conflict and disagreement. There can be different and multiple sources of conflict.

Personalities

Different individuals may have conflicting personalities. You cannot change personality type. Both people could be crucial to the project. Hiding or not recognizing the conflict just drives the problem underground and makes it worse. There are several possible approaches. One is to meet with the two people at one time and indicate that you are aware of the issue and that you do not intend to change anyone. You are just acting in the best interest of the project. Try to appeal to them on that basis. Another approach is to talk to each person individually and get their feelings out. Then attempt to reconcile the differences or try not to bring out the conflict. This has a lower likelihood of working. With personality differences in a project meeeting, try to keep meetings focused on issues.

Money and Resources

In a larger project, you can have conflict between different project leaders as they compete for people or other resources. You as the overall project manager have to be in the middle. This should be deflected in advance through the project plan and the allocation of resources. If requirements and the critical path changes, then you may have to realign resources.

Priorities

A person in the project team might want to work on task X, but task Y is more important. They continue to spend time on X when they should be spending time on Y. What do you do? You should meet with the person to go over the plan. Look at what they are doing on X and Y. Do not continually tell them not to work on X. This may just drive the work underground. Divide up Y into small parts. Have them show you what they have done on individual parts frequently. This can make a difficult or unpleasant task more palatable. Why do individuals do this? They are human; they like to work on things for which they are trained and which match their skills.

Procedures, Methods, and Tools

Different people have different concepts and ideas about the best way to carry out a method or use a tool. This can be based on their previous experience in other projects. It can also be based on fear of the unknown. Some team members may view doing the work and learning something new as too much to do. They may have no direct knowledge, but have heard that there were problems in another project. In some cases, this is not a problem. If the result is acceptable and the method is generally followed, there can be variation. The project manager establishes guidelines on how the procedure, method, or tool is to be used in the project. These guidelines should focus on the minimum that will be necessary. Asking for a lot more may retard the project progress. We know of a manager who enforced every rule to the letter. People followed these rules and the project suffered. A project manager should "lighten up," but not cave in to pressure. A key suggestion is that the guidelines imposed by the manager should be consistent across the project and across time.

If certain individuals are exempted, the procedure or method is jeopardized and threatened. Another part of the strategy is to get the differences out on the table early in the project as you are setting up the guidelines. You would lay out the method or tool and ask for suggestions from the project team on guidelines. You are concerned with effective and efficient implementation. You also care about verification and validation. It is one thing to suggest that people use a tool; it is another not to have specific approaches for verifying the use of the method or tool. A suggestion is to have someone in the project team act as an expert to whom the team members can go for assistance. You can also invite people outside of the project who have used the method or tool to address the team and answer questions. Do not rely on people asking questions. Get questions individually from project team members prior to the meeting and ask them yourself. This will cover the team in terms of anonymity.

Schedules

People may disagree on whether they can do the work. At the start we indicated that team members should be involved in defining schedules. This will help, but conflict can still occur. Like other items, treat a schedule question as a symptom. Try to find out what is the underlying issue. Try to do this individually with the team member. It will not do much good to the project to discuss this in the group. People will be attracted to the issue and the conversation will get beyond the original scope.

We have learned that there are several ways that people handle conflict. Some people withdraw. They do not want confrontation. They sit in the project meetings and say nothing. But they are not buying into what is being decided. You should work with team members individually prior to a meeting. Identify people who might withdraw and talk to them. You should recognize that this is a common reaction. A second way people deal with conflict is to paper it over, deemphasize the problem or issue. This may get everyone through the meeting, but the issue still is there. You should address it in the context of the issue. Recognize that there are different points of view.

We could also have two opposing sides in a conflict argue it out. This gets the different view out in the open. Here you would meet with each side and present the other side's position. Then when the meeting occurs, you could point out the differences. Do not openly take one side or another unless the project is threatened. Your position as a project manager is to get the project through to a successful conclusion. There are many ways to get there.

Compromise is a fourth method. Unfortunately, there is sometimes little ground for compromise if the positions are very different. There are dimensions of compromise. You can compromise on tactics and details. You cannot compromise on the strategy and the objective. That was set at the start of the project.

A tool that has been useful in heading off conflict is the use of groupware. Groupware allows the team to share information on issues and to have updated comments. Comments can be analyzed quickly; there is more rapid feedback.

RESULTS OF USING TOOLS AND METHODS ARE NOT MEETING EXPECTATIONS

The tools or methods are being used, but they are not producing the desired result. What do you do? First, try to determine how the method or tool is being used. Do not assume that it is being employed incorrectly. Second, you might conduct a review of the use of the method or tool within the project. What are some of the factors behind this? One possible issue is that the method or tool is not a good fit with the project; it is a forced fit. A second possible issue is that people lack experience in using the tool and need help. You should treat this as positive with the goal to improve use of the method or tool, or modify guidelines in its use. Don't be willing to jettison the method or tool too easily. You might end up with chaos as people do anything in any way. The assumption is that the method or tool was selected for good reasons. The problem lies in its application in the project.

This emphasizes the need for measuring the effectiveness of the method or tool. You should also conduct periodic reviews of all methods and tools and solicit suggestions based on the experience of the project team. This can be a very positive method for getting better guidelines. People have a growing role in their application.

There is the other side of the coin. People can be in love with the method or tool, but it becomes evident that there is a mismatch between the method or tool and the project. Here you would work on better use. But you might start looking for alternatives to the method or tool. Changing a method or tool has risk, but there is risk in continuing failure. Let's suppose that you have a new tool that you would like to try. What do you do? Sit down with some tool expert and find out about transition and implementation. How long does it take to learn? How long does it take to become proficient? How can its use be measured? These questions will help you decide if you want to pursue it further. If you do, then you might surface it individually to project team members. You might find a volunteer to use it on a pilot basis. We point out that we, in general, do not advocate switching tools or methods in the middle of the project due to its disruptive effect on the project.

This discussion supports the concept of starting the project with as few tools and methods as possible. It will be easier to introduce new ones in different phases of the project. Keep the tools in later phases vague and flexible. Introducing something new when there is no set tool in use is better than replacement.

PROJECT TEAM MEMBERS ARE LOBBYING FOR CHANGE

People in the project team mention that they are dissatisfied with how the project is going. They may be challenging the manager of the project. You could treat this defensively and just say that is the way it is and you are the manager. In the 1990s with accountability, this will not go far. You have to channel this into positive change. First, you should get a list of issues out on the table. Second, you are trying to understand the range, scope, and content of the issues. Don't assume that they are trying to undermine the project, although this is a tempting thought. Turn it into a positive thought. They are concerned about the project because they want it to succeed. If it succeeds, they win. Be open to suggestions as to what can be done. Map the suggestions to the issues. These issues may fall under some of the other symptoms we have discussed. Develop a straw man set of changes and see if these will address their concerns. Address each member of the project team with the straw man.

POTENTIAL CAUSES AND ISSUES

Underneath these symptoms, there can be a variety of issues. We will address some of the common ones here.

Project Not Recognized as a Project

Some line managers do not recognize the work as a project even though it has management support. The symptoms of this include denial of support, lack of resources, and other bottlenecks. You could go to management for support; you could confront the line managers. Or, you could show that the project is real by showing results. This is better. But you still may have to go to the line manager on specific resource issues.

Too Many Simultaneous Projects

Your project manager is torn in competing directions. Management may not want to favor one project over another for political or other reasons. Now is the time that you should focus on a core set of tasks to illustrate results. If you absolutely require a specific issue or resource problem to be addressed, then go forward. Do not press a case that is general and vague, because your abilities as a project manager may be questioned. In the current decade you must make do with less.

Impossible Schedule Commitments

Like the preceding point, complaining to management is not an effective strategy. You have to make your stand and adhere to a specific issue and a set of tasks.

Poor Project Manager

As a team member you might try to work with the project manager and improve the situation. You might volunteer to work on other tasks. If this doesn't work, then you should wait until a milestone is not met or an important issue surfaces. The issue or milestone slippage should be the catalyst—not you. The fault may not be with the project manager. It could be with the structure of the project or with other factors of which you have no knowledge. The project manager has more contact with management than you. They may know more about other factors.

Poor Control over Changes

The project seems to be out of control. Tasks and directions shift more than the wind. In this situation, a team member needs to focus on the strategy and objective of the project. Change and chaos can be going on around you, but concentrate on the work. If you are the project manager and you see this happening, do not overreact. Do not freeze changes. This is an overreaction. Move to an approach where you handle several changes at one time—not just one at a time.

Problems in the Project Team

There can be any number of issues involving the team. Some team members may lack experience. You might pair up junior members with senior members to accelerate the experience. Some team members may not be competent in certain tasks. You can give them help yourself or have another person help them. If these steps don't work, then you might have to change task assignments. Do this as part of other changes. Otherwise, people will get discouraged and productivity will drop. Team members can get immersed in side issues and trivia. Although some of this is natural, keep the attention on the tasks and the management critical path. One technique we have used is to have a trivia time in the project meetings. This allows people to vent their frustrations.

ISSUES IN MULTIPLE PROJECTS

Conflicts and issues can arise among multiple projects in several areas. One is that the projects have interdependent tasks and milestones. To head off problems here, build an overall summary schedule that includes all interdependent problems. Track this megaproject along with the independent project so that when changes are made to the individual projects, they are rolled up into the megaproject. This will support issues being rolled up to a higher level.

The second area is overlapping resources. Here independent project managers are competing for same facilities, equipment, or personnel. Knowing this from assessing the projects, we can construct a schedule for each such resource by extracting tasks from the individual schedules and plans that use the resource. This can be set up to be created automatically. Then the resource can be allocated. The allocation formula can be driven down to the individual projects.

The network should be used to share information about conflicts and issues between projects as well as for opportunities where one project can assist another. Consider having a group meeting of project leaders in which the leaders work collaboratively to do what if analysis. Use a projector that can show the computer GANTT chart on the screen. Show a view that contains the GANTT chart on the top and a histogram of resource allocation on the bottom. As you change the allocation in the GANTT chart, the resource graph will smooth out. A final lesson learned—don't break up the meeting without a deal.

EXERCISES

The first thing we suggest is to take any project and make two lists. One is for the symptoms and the other is for the issues. Try to match these up. This will give you a better understanding of the differences between symptoms and issues. Take the same project or another and ask yourself about the process of change that was carried out in the project. Were the changes clearly related to issues? Did people know why the changes were being made? Asking questions such as these shows you how change and action relate to issues.

EXAMPLE

What might be interesting here is to highlight some of the issues that arose in the project and also methods used to resolve conflict. One issue that arose was the technical response time performance of the network project management system and database. To resolve this, an in-depth, lengthy academic investigation could have been undertaken. This would have taken too long and delayed the project. Instead, several possible countermeasures were identified including tuning the system, upgrading the hardware, and adding better software. Rather than trying these one at a time, all three were implemented in parallel. The problem was solved, but to this day, we do not have a nice theoretical reason for the contribution of each measure to the solution.

A second issue was the resistance of one division to participating in the project even after demonstration of benefits and effect in final assembly and integration. The decision was made to bypass the division and move to other divisions. This was coupled with a strategy of surrounding the recalcitrant division so as to exert more pressure. This worked and they came on board later.

A third issue was the underground resistance of those project management support staff who supported the old system or *ancien regime*. Rather than ignore this issue, the decision was made to address it head-on. Training was initiated and several staff members were temporarily added to the project team. In addition, the role of the support staff was redefined in the context of the new project management method.

Another issue was success. The project expenditure rate was less than expected. A manager threatened to take some of the money for other projects. To head this off the first time, the expenditure run rate was accelerated. On an ongoing basis, the run rate was monitored constantly. Success has strange side effects at times.

To address conflicts, an early effort was made to encourage extensive discussions of issues. This was done in the context of the project management system. Issues were tracked and each side of a conflict was reported as a legitimate point of view. The project team worked to keep aloof because they were concerned with implementation of change as opposed to how a detailed reengineering issue turned out. Our focus was to get resolution and keep it resolved. Conflicts that were resolved would be actively revisited in meetings to reinforce conflict solution.

Groupware was extensively used to track issues and to share comments. Proceeding from a slow start, the approach has now been adopted to identify and track conflicts and issues long after the project was completed.

SUMMARY

We have dealt with some of the common symptoms and issues that are encountered in projects. But remember that each new project will surface new issues or new nuances of old issues. Constant through all projects is the need to differentiate symptoms from issues. By first understanding the symptoms and then the issues, you have a chance to turn the situation around to an opportunity. You should view issues as a way to improve the project. There are very positive effects to the surfacing of an issue. It shows that people care about the project. Also, it gives you the ability to make changes. If everything were going fine, it would be difficult to get changes through and into implementation. Changes in even small projects are natural. As time goes by, you learn more about the project. You learn more about how organizations are dealing and interfacing with the project. You also see more clearly the effectiveness of the team, the methods, and the tools. One can make the case that if a project has no issues, then the project is of little importance to the organization and that people do not care enough about the project.

IV PROJECT MANAGEMENT TOOLS AND METHODS

14 Project Communications and Presentations

Project communications include the medium or the way we communicate, including the following:

- In-person informal communications
- In-person formal meeting
- Telephone voice mail
- Telephone contact with receiver
- Telephone contact with human being, not the receiver
- Casual note
- Informal memorandum
- Formal letter or document
- Electronic mail
- Web pages through the Internet or intranets
- Videoconferencing via the Internet

Within some of these there are additional options related to location. We can consciously control which of these we want to use in a project. By considering the time of day, you can predict with some degree of certainty whether the receiver will be present and the degree to which they can focus on what you are saying through your message via the medium.

ADVANTAGES AND DISADVANTAGES OF DIFFERENT MEDIUMS

Table 14.1 presents some advantages and disadvantages of some of the different mediums. Note that there are advantages to verbal communica-

TABLE 14.1   Advantages and Disadvantages of Different Mediums

A. Telephone call
1. Disruptive to the person receiving the call unless planned in advance
2. Likelihood of getting through less than 30%
3. Needs to be short and directed/focused
4. If you leave a message with an individual, make it very short.

B. Voice mail
1. Useful for short messages
2. Needs to be structured—who, reason for call, what is needed
3. Can be misused by leaving long messages

C. Electronic mail
1. Provides a written electronic record
2. Can be rerouted by receiver
3. Sender usually knows if mail was read
4. Requires precise title so that person knows subject before opening the mail
5. Effectively limited to one screen in length because that is what can be conveniently read
6. Cannot use with sensitive matters

D. In-person visit
1. Perhaps the best for one-on-one meetings between project manager and team member
2. Can get more complete impression because can see person in their own working environment
3. Can be informal and cover wider ground
4. Needs to be focused and not just a rambling meeting

E. Informal note
1. Friendly way to communicate in this age of electronics
2. Message must be clear and concise

F. Formal memorandum
1. Should be restricted to very specific and important topics
2. Can be misunderstood and result in more effort to clear up the confusion

G. Web pages
1. Easy to establish
2. Static; must be updated to maintain interest

tions in the sense that you can convey tone and yet not be pinned down to specific text. In project management verbal communications cannot be underestimated. For the same reasons, electronic mail is fraught with risk. Not only does someone have your written message, but it is an electronic form that can be forwarded and copied to anyone. It is much more credible than other written communications because it tends to have been written by the sender without a secretary. Web pages are used for showing project reports in color as well as graphics. The key to their use is to maintain them. Looking at a web page that hasn't changed in months can really turn you off on projects.

MESSAGES ARE GOLDEN BULLETS—USE THEM SPARINGLY

In general, people communicate too much. With more options and time compression, you should view messages as golden bullets—fire very few, but make them count. Your messages will get more attention. Here are some guidelines in preparing your message:

1. Think about the message and the cascading effects of what will happen after the message is sent. This will impact your other decisions later.

2. Decide who is the receiver of your message. This is not just the main receiver, but also people who are copied (cc) and given just the cover sheet (blind carbon copy—bcc). Think it through politically.

3. Decide on the medium and timing of your message. We have discussed the medium and will cover it more later. Timing is another issue we will discuss.

4. Package the message adhering to the guidelines imposed by the medium that you have selected.

5. Separate the message from the structure and these in turn from the grammar.

6. Follow up on your message to see how effective it was.

FIGURING OUT THE AUDIENCE

At the start of this process, you think about the person to whom you want to send the message. You also have a general message in mind. Now sit back and think. If you contact this person and convey the message, what will happen? Will the person tell other people, such as their line manager? Because you are going to send a message, you should attempt to accomplish several purposes with the same effort as long as you do not cloud the message.

This thought process will cause you to potentially expand the message and adjust who will be the receiver of the message. Consider also the various classes of receivers. The people to whom the message is addressed are expected to respond. The people who are copied (cc'd) are given the same information. You would likely include people who will have to support the ones that receive the message directly. On the other hand, there are managers and others who need to know what has been requested or what issue has been raised. They need to receive the cover memo and not the details. This is the reason for the blind carbon copy (bcc).

FORMULATING THE MESSAGE

You first have to write down an outline of the message you want to send. This is not the same as the format that will eventually apply to the message. The latter will be based on the medium used. How do we formulate a message? We need to give an introduction. We have to identify the question, issue, problem, opportunity, or whatever. The message has to be placed in context. We also have to convey the expectations that are attached to the message. Expectations can be stated in terms of what type of response is needed and when it is needed. To further support the message you should indicate how the receiver's response and work will be used. We have developed and used the form in Figure 14.1 In this form you will identify the audience by type at the top. You will then give the purpose of the message.

```
                             Message Form

Audience: _____person/group_____   Date: _____

Purpose: To _____why are you doing this_____

          _____

Action: _____what will happen_____

          _____

Expectation: ___what you hope to achieve_____

          _____

Medium: _____   Timing: ___when_____

Title : _____must be clear and short__
```

FIGURE 14.1 Message structure form. The purpose of this form is to encourage you to spend more time structuring project-related messages. This will help to focus your thoughts prior to telephone calls or meetings.

The middle part contains the outline of the issue, the action expected, and expectations. At the bottom is the medium selection, timing, and title of the message.

Figure 14.2 gives a single message that can be conveyed using one of three mediums. We now need to consider different guidelines for alternative mediums. Then we can select among the types of medium and determine timing.

ELECTRONIC MAIL GUIDELINES

Electronic mail is software that typically allows you to compose messages, send messages to groups of people, attach files to your message, and perform forwarding, printing, and other functions. A more detailed list of electronic mail features is shown in Table 14.2 along with a set of comments as to how this feature could be used. Although the features of the electronic mail system impact your selection of medium, a more central concern is the audience that is served by the same electronic mail system. If only 5% of your audience is on the electronic mail system, it may not make much sense to use it.

Now let us go to the guidelines:

- Do you know how to use the electronic mail system? You should practice by sending messages to yourself. If you do not know how to use the system, you will likely make a simple mistake that the receiver will notice and then think less of you.
- Be careful with spelling, punctuation, and capitalization. Most electronic mail systems cannot check for this. You are on your own. It is very embarrassing for a manager to show ignorance of English grammar rules. This surfaces when a manager makes continual errors. It shows that either the sender does not care or that he or she does not know any better. Either way, the sender loses.
- Set up a group name for the inner and outer project teams. An electronic mail group is a set of people who can be addressed under a group name—say project A. The inner group consists of the people who are regular players. The outer group consists of people who are involved in the project in any way and who should be kept up to date with general information.
- Be very careful in the phrasing of the title of the message. Convey the meaning, but not the action. For example, suppose that we wanted to have someone take action on issue 43 relating to the use of an assembly tool. We could name the tool and title, but we only have 50–60 characters in the title. Instead, use the title "Issue 43 Possible Action."
- Keep the message short. Let's suppose that you need to convey substantial information—more than is visible on one screen. What you need

A.   In Person Meeting to Determine Status of Team Member

                        Message Form

Audience: John Doe, Engineer_____   Date: _XX/XX/XX_

Purpose: To determine status and see how use of tool in_____

         computer aided design is going._____

Action: _Status, determine adequacy of documentation of tool

         _____

Expectation: ___Determine any issues and support needed_____

         _____

Medium: _____In person meeting_____   Timing: 10 AM-20 min

Title : __John Doe Status/Use of CAD__

B.   Electronic Mail Message to Project Team on Next Meeting

                        Message Form

Audience: Project Team_____   Date: _XX/XX/XX_

Purpose: To provide agenda and solicit opinions on issues___

         prior to meeting_____

Action: _Meeting details/agenda/3 key issues identified_____

         _____

Expectation: ___Get feedback on issues prior to meeting and_

         confirm everyone can attend_____

Medium: _____Electronic Mail_____   Timing: 8 AM-1 screen

Title : __Project Team Meeting- 4/3- 9:00- Agenda/Issues

FIGURE 14.2   Examples of messages.

to do is to create the document as an attachment document. Then
the electronic mail message is a cover sheet or yellow stickie message
that refers to the attachment.

• Always copy yourself. This sounds truly stupid. However, because of
  limitations in most electronic mail systems, you do not automatically

C.   Voice Mail Message to Manager on Issue

- -
Message Form

Audience: Mary Smith, Manager_____   Date: XX/XX/XX

Purpose: To inform manager of issue status and resolution___

on Issue 38_____

Action: _Elicit response and reaction to what is being done

_____

Expectation: Hope for call back later in day affirming___

support_____

Medium: Voice Mail_____   Timing: 9 AM-2_min__

Title : __Mary Smith- Issue 38_____

FIGURE 14.2   (*Continued*)

get a copy of what you send. As a habit, you should send a copy to yourself.

- Use the pending function. Pending is a nice option in systems. You can select pending and see who has not opened the message you sent.
- Storing and printing electronic mail. Electronic mail systems archive and purge messages on a periodic basis. We suggest that you save the electronic mail on your computer disk.

TABLE 14.2   Electronic Mail Features

Ability to attach spreadsheets, documents, project plans, graphs, or other files to your electronic mail message

Length of title of message is at least 50–60 characters long

Linkage to electronic fax so that you can send and receive facsimile messages with the same system

Ability to have electronic forms within the software

Ability to attach fax to electronic mail message for routing

Ability to copy and blind copy (cc and bcc, respectively)

Ability to create, modify, and delete mail groups

Ability to set priorities for different types of mail messages

Standard interface to external electronic mail using international and other standards (e.g., ANSI X.400)

Ability to have system beep at you when you receive mail of a certain priority

Ability to access mail from remote locations with the proper security

- As a receiver, you should join as many groups in the organization as possible. This will inform you about activities and functions in the company. This is very useful and it does not cost anything. Most systems allow you to tag the items you do not want based on their titles and then delete them in a batch mode.
- As a receiver, begin to be critical of the messages you receive. Ask yourself how the messages could be clearer and more precise. This will help your own skills. Print out some good and bad examples.
- When you receive a message, do not immediately respond. This shows that you are in a reactive mode and respond on the spur of the moment.
- When you respond to a message, decide whether you want the sender's message to be embedded in your message. If you want to get your point across, then do not include their original message.
- In writing and responding to messages use complete sentences. There is often a tendency to abbreviate as some people associate electronic mail with the telephone or telegraph. Treat it like a letter.
- Use Internet and other global electronic mail systems. These are very useful for international communications on projects. We have used this capability extensively. The same guidelines apply.
- You can use electronic mail as a common place for team members to comment on a particular topic. They might provide suggestions on a specific tool, for example. Here is what you can do. Set up a document entitled the tool name or some other descriptive passage. Then when you access it you can update it and sent it to the group.
- Develop a set of canned electronic mail formats and messages related to typical project activities. Make these available to the project team so that they know what to expect. Also, encourage them to adopt standard formats.

Electronic mail is a powerful tool. We will see how and why you want to use it for status reports and for sharing information on issues in the project. It is a way to get to people more quickly and with assurance that the message was received.

GROUPWARE

Groupware allows people to share information and collaborate on projects and resolve issues. Groupware and web software have overlapping features. Databases are stored for common access across the organization. Benefits of groupware to projects include (1) improved communications; (2) ability to structure information sharing; (3) greater ability to organize information over electronic mail; (4) support of getting consensus on issues. Groupware can allow to set up forms that relate to the databases. Lotus Notes is an example of groupware. As an example, you can create a form for recording action items using groupware or web software easily. Information input can be saved and shared among staff. With a web browser approach we avoid installing special software on each PC—giving an edge to the browser and a cost savings for project management.

VOICE MAIL GUIDELINES

Voice mail is becoming more prevalent. The cost of receptionists and secretaries along with the declining cost of the technology make it more popular. Because it is so widespread, you should probably use it in projects. Here are some specific guidelines.

- You should record your greeting and listen to it several times. How was the pace? Did you sound nervous? Is there any background noise? There are people who change their message every time something new comes up. We recommend against this because it shows you must have a lot of time available to do the changes. You also are telling people about yourself—maybe you do not want them to know. Tape one message and stick to it.
- Should you leave your voice mail on when you are in the office? This one can be argued either way. On the one hand, if you want to show good service, you should answer the telephone. If you want to be organized, then use the voice mail to screen the incoming calls. It will give you more time to think about your response.
- Should you use call forwarding? Only if no one else is using that telephone. Otherwise, the caller has yet another step to go through.
- How should you listen to your messages? A problem with voice mail is that when you start to listen, you are only told how many waiting messages you have. You have no idea who called. Let's suppose you are busy and the phone time is short. People are waiting behind you in line. Or, you are in a open area. What you should do is start listening to the messages. Develop the habit to listen to the first 10–15 seconds of the call and then either fast forward and save it or kill it. Do this for all messages.
- How should you send a message? As with the electronic mail, you need to think about the message ahead of time. When you call, you should expect voice mail over 40% of the time. Your message has to be formulated in three forms: you reach the person; you reach a human to record the message; you reach voice mail. We will consider the last two here. If you reach a human other than the receiver, identify yourself. Always smile into the telephone.
- Tell them your name and telephone number and a very brief message. Do not leave a message that you called without identifying the topic. The topic should be in two or three words. Do not give a time for

them to call—this will look like you are trying to control them. For voice mail, you should start with your name and the purpose of the call. Do not give the date and time unless you know it is an answering machine. Look at your watch or have a timer handy. Keep the message to under one minute.

- You should get back to people who have left a message with an acknowledgment of their call and what you plan to do in response.

FACSIMILE GUIDELINES

Many of us wonder how we could have ever gotten along without fax machines. They are a wonder because they not only convey written information but signatures as well. The original fax machine was developed in the 1800s. People at the time wondered who would want to use it because it would disrupt the work of the day. With daily mail, you generally know when the mail is going to reach you. You can plan your day around the receipt of the mail. It is organized. Now turn to the fax machine. It tends to be disruptive. But for many businesses, fax machines are critical. They are very widely used in international business. Now for some guidelines.

- Design a cover sheet that allows you to place text in the bottom half of the page.
- Select the time of day for faxing carefully. Some offices are very busy at certain times. Your fax could end up on the fax room floor.
- Call the person and tell them you are sending a fax right after you have done it and after you received a successfully transmitted code. If you call in advance and the fax does not go through, then you get to play telephone tag.
- Do not assume that the fax machine or location is confidential. Fax is much more public than voice or electronic mail and so should be used sparingly for sensitive or confidential topics.

In projects we would typically use fax for notifying people of meetings and agendas for meetings. We would not send results or questions via fax. We would not address issues by fax.

MEMORANDUM AND LETTER GUIDELINES

This is not a book on how to write letters. But we can give some suggestions on structure and content as they relate to projects.

- You should write the letter or memorandum yourself first. This gives you more control over it. If you give it to someone to type and after they type it, you want some changes, you will feel bad about asking for changes.

- Before you start writing, make some notes and an outline on a piece of paper in handwriting. You have to get in the habit of being organized. Ultimately, you want to be able to compose memoranda on-line with your word processor with no outline. Just sit down and do it. Very few of us ever get to that point and do it well. Structure your letter into paragraphs. The first paragraph should state the issue or topic of the letter. The second paragraph can give some background and detail. The last paragraph should identify the actions that you expect and what will happen next.
- Do your own distribution. That is, make your own copies and distribute these by hand. This gives you an excuse to meet with people on the project team to review the results.
- Develop standard formats for specific project activities. Some examples are (1) notification of project change; (2) notification of meeting; (3) meeting results; (4) problem identification; (5) problem resolution.

WEB PAGES

We consider three aspects: deciding on content, preparing the page, and maintaining the page. The goal of content is to provide both static and dynamic information for use by the project team, and secondly, by managers and staff outside of the team. Static information might include the description of the project, the project team, goals of the project, the scope of the project, and milestones of the project. Dynamic information can encompass status, GANTT charts, recent milestones, lessons learned, issues, and action items. What you select depends upon what you want people to see and do, and what you are willing to maintain. It is estimated that half of the web pages are never maintained after being initially established. Another question is "What data are you willing to display if the project suffers a reverse?" Keep the content consistent. Otherwise, people may sense problems if the content suddenly changes.

Generating the material is the first step in preparing the web site. Carefully think about what graphics you wish to display. Some possibilities are: pictures of team members, a picture of the end result of the project, the location of the project, and the organization. The second step is composing, integrating, and publishing the page. Get a page up and running fast so that you understand the process. Don't show this to anyone. Refine the page and get feedback from the team. Cover the update method and how they are expected to use the page. After all, they are the main audience and will be providing updates and using the information. After you have concurrence, publish the page. Get the team used to the page and refine it more. After the page is stable and in use, advertise the page to management.

Maintaining the site is the major job. While team members provide some detailed information, the project leader has to ensure that the major parts of the page are updated and consistent. How often should the page be updated? Try one or two month intervals. If it is more often than this, the workload is too great. If it is less often, the page is too static and dated.

VIDEOCONFERENCING

Doing videoconferencing on the Internet requires a camera, microphone, speakers, computer card, and software. Each site must have compatible hardware and software. Carefully select a site based on space, activity and the lack of distractions, and lighting. The quality of the video is limited by several factors, including the speed of the Internet itself. A key factor is the number of frames per minute, that is, the refresh rate. The more frames per minute the more communications capacity is needed; the lower the number of frames the jerkier the image. It is a trade-off. Even so, the results will not be smooth video.

What are some of the applications of videoconferencing? Here are some examples.

- Discussion of drawings, charts, and other tables where visual displays are useful
- Discussion of issues and action items
- Technical discussion about specific project subjects

We prefer not to employ it to obtain status. This can be done by telephone, voice mail, or electronic mail.

Test out all software and hardware by using the technology for general discussions. Try a conference call with several sites. Get people comfortable with movement, body language, clothes, placing displays and exhibits, and voice. Establish the schedule for videoconferencing through e-mail.

INFORMAL CONTACT GUIDELINES

By informal contacts we mean instances where you meet with someone on the project team outside of a meeting. Why do you want to do this on a regular basis? People will likely talk more when they are away from the meeting. We suggest that you visit them in their offices. Do not have them come to you even if you are the project manager. Show that you respect them by visiting them.

Do not always preannounce your visits. Stop by when you are in the neighborhood and set up a time for a later meeting. A person's office tells you a lot about the person. To what extent has the office been personalized? What papers and documents are out on the desk? Are they about your project? If you visit them four different times in a month and your project is never out, then you can wonder how much time they are spending on it.

Begin with general questions as to how their tasks are going. Ask about their other work. If this part of the conversation produces some leads, follow them up. If it does not, go to methods and tools of the project as well as to issues. Ask them what you could do to make their life easier or better. Is there anything they need? Are they having any problems? When you meet with people informally, do not take a pad of paper. If you have to take notes, ask them for paper. When you make notes, tell them what you are writing and how you are going to follow up. Then make sure and follow up.

For economies of scale try to visit several people in succession. If they are not in their office, leave them a note that you dropped by to say hello. You should attempt to get around to people several times per week regardless of the size of the project.

COMMENTS ON FORMAL COMMUNICATIONS

Projects have a wide variety of formal communications in terms of meetings and paper documents. For meetings we encourage you to consult the project team or meeting attendees with the tentative agenda for the meeting. Get their input and resolve their questions. Circulate the final agenda. At the meeting follow the agenda. In defining the agenda, we like an approach where the mundane items are handled first. We put status reports at the start. This leaves most of the time for issues. Project meetings should, in general, focus on issues and opportunities.

Here are some specific formal, project-oriented meetings that you will encounter:

- Management meeting—delivering either status or addressing an issue.
- Coordination meeting with other projects—This needs to be formal due to the interfaces between systems and projects.
- Kick-off meeting—This is a formal meeting because it needs to stress organization.
- Milestone review meeting—This meeting should be formal and follow a specific checklist. People who reviewed the milestone should have given you materials in advance of the meeting.

DECIDING ON THE MEDIUM AND TIMING

Here are some suggestions for selecting the medium:

1. Keep as much of the project issue discussion to in-person and telephone communications. Avoid the written word because people may be reluctant to commit to positions on issues in writing.

2. Use electronic mail as much as possible for the routine and mundane work on the project. When distributing reports, send people an electronic mail message saying that it is coming.

3. In general, avoid facsimile because it is too open and available. Also, much of the project work is important but not immediately urgent.

4. Have as few formal meetings as possible. There are several reasons for this. First, the meetings will be better attended because there are fewer meetings. Second, have the meetings focus on issues and opportunities and not on status.

5. Test videoconferencing for your project if the team members can access compatible technology and are spread out geographically.

Try to do as much as possible off-line when only a few people have to be involved.

WHOSE FAULT IS IT IF THE MESSAGE FAILS?

What is failure? Failure here is taken widely. If any problem occurs, it is a failure. You have to hold high standards. Examples of failure are

- Message received not at all or too late.
- Message was misinterpreted.
- Message was sent to the wrong person.
- Message was ignored.

Who is responsible for failure? You are, as the sender. The sender framed the message, selected the timing, and picked the medium. The sender can make reasonable assumptions as to how the receiver will interpret the message. Therefore, it is the sender's fault. The motto here is to be hard on yourself to improve yourself and yet be kind and forgiving to others. Of these four problems, concentrate on the misinterpretation issue. It is the one that can be most harmful.

EXAMPLE—PROBLEM NOTIFICATION MEMO AND PROBLEM CLOSEOUT MEMO

Whether by paper or electronic mail, you need to inform people about problems and their resolution. In other chapters we discuss details of what should be in these messages. Here we discuss format and structure. Some common communications problems are as follows:

- Problem definition is never clear.
- How the problem was resolved does not match the definition of the problem.
- Issues are not tracked and managed; the same issue recurs.

Although we have made specific suggestions elsewhere, we want to emphasize that the project team needs to be aware that the relationship between the issues and actions can be many to many. In handling issues and problems, we urge you to consider electronic mail in solicit opinions and share information about how to resolve issues without spending additional resources. This communications process will encourage greater understanding and kinship among the people on the project team.

TESTING YOURSELF AND EXERCISES

Figure 14.3 gives some project examples that tie the above discussion together. You should also review what you have done in the recent past. What other options could you have considered as opposed to the medium you chose? What about timing? Did you select the appropriate time? How do you sound?

PRESENTATIONS

We end this chapter by considering the specific area of project presentations. There are several specific areas where you need to make decisions

1. Format—medium of the message. An example is type of memorandum with copies and so forth.
2. Length—how long will the message be? This is dependent on the subject and the receiver's interest.
3. Organization—what to start and end with and the order of presentation.
4. Method of argument—do you use project data, your experience or authority, or the history of the project?
5. Your attitude toward the receiver(s)—formal, polite, informal, friendly—you decide.
6. Impression you make—what should the receiver think of you?

You have the strategy defined. What now? First, you need to assemble the materials you are going to convey. This can be graphs, tables, and so forth. This is the evidence you are going to use—the hard data that supports your position or statements. Second, write down an outline of the order of presentation. In project management you should be able to do this in ten points or less. Allocate the space (in your strategy) to each of the points in your outline.

We have talked about format and structure. We have three choices in project communications.

1. Electronic Mail to Project Team on Issue 37

    To: Project Team
    Subject: Iss 37 Status/Possible Actions- Need Your Input
    Message: Issue 37 on general ledger chart of accounts-
    Alternatives have been defined. 1) Keep the expense
    categories as in the current chart of accts.; 2) use
    more detailed list from Mary C. Issue is whether detail
    in alt. 2) can be captured. Any ideas?

    Note that this message fits on one screen and is short
    and to the point. Topic, description, status, and
    request are all given.

2. Voice Mail to Manager Related to Issue 37

    Hi, John, this is Bill Strong. On Issue 37 we are
    favoring Mary C.'s idea for more detailed expense
    information in the chart of accounts. Question is
    whether this is feasible. Hope to hear from team
    tomorrow. Any thoughts? Otherwise, I will get back to
    you with results.

    Note that this is short and the default is that the
    manager has to take no action unless he has an idea.
    This makes it easier on the manager and keeps them in the
    "loop".

3. Memorandum on Project Changes

    To: Project Team

    From: Bill Strong

    Subject: Changes to Project Plan, Effective May 1

    The attached task descriptions, GANTT chart, and PERT
    chart summarize the changes to the project as a result of
    expanding the project scope to include the two additional
    sales locations for our new sales management system.

    These changes are consistent with our last meeting.
    There are issues in regard to:

    o Scheduling the training of the additional offices
       with the same staff
    o Getting the managers in the two locations
       involved in the project

    Please review the attached and respond with any questions
    and comments on the above issues by April 14.

Note that this example describes the change very briefly,
leaving details to attachments. Specific actions are
requested of the project team by a specific date.

FIGURE 14.3 Examples of project management messages.

TABLE 14.3   The Project Presentation to Management

The project manager opens the presentation by discussing the current status of the
  project. This parlays into a discussion of specific problems and issues within the project.
  These issues are addressed and the meeting ends.
What went wrong? There was no overview to place the current state of the project in the
  context of the history of the project. Management has to review many projects. They
  cannot be expected to remember each and every project, much less detail within a
specific project.
  Some specific criticisms are
  1. Format—The format is bottom-up and chaotic.
  2. Organizations—The presentation is disorganized and lacks structure.
  3. Direction—There appears to be no beginning or end—just miscellanous issues.
  4. Feelings toward Audience—The project manager is assuming that management is up-
     to-date on the project and envisions the project and the manager's role clearly.
  5. Impression—The project manager can appear arrogant or condescending.

1. You can describe a situation, status, or issue. This is passive. Don't
expect a lot of attention. The audience will wonder why you are discussing
it. What is it leading up to?

2. You are writing about some analysis results. This is typically the result
of doing a postimplementation review, a review of alternative project approaches, or the results of reviewing a milestone.

3. You are writing to persuade. This is just what it says. It is the underlying
reason why many of us write.

EXAMPLE: DESCRIPTIVE REPORT

You first have to introduce the subject. Also, identify the reason for
doing this. In this opening paragraph, you may also have to give your
qualifications—establish your credibility. Isn't this obvious in a project?
No. How does anyone know that you are an expert in some method, tool,
or activity—especially if you are not doing it. Give a statement as to what
you reviewed and did. Next, give an overview of the topic. For example,
address an issue in general with a focus on the impact and different perspectives on the issue. The third step is to plunge into details. Remember to
structure these details in a way that will not confuse the audience. The last
step is to bring the audience back to the issue as a whole. You are putting
together the parts you described.

EXAMPLE: ANALYSIS REPORT

Here we refer to problem solving, issue identification, justification for
actions, procedures, and technical analysis. We begin with an identification

of what we are analyzing. Like descriptive reports, we have to establish our credibility and expertise. The current state of affairs that was the subject of the analysis needs to be described. You can use the general, specific, and general method presented above. Define the methods you have employed in the analysis. Project management can depend on interviews, review of files, testing results, design and specifications, statistical analysis, and so forth. You should also lay out the order of your analysis. Where will you begin? Where will you end? Give the analysis results or findings. This should exactly match the rules you have just given. Finally, what does the analysis show? Where do you go from here? Here is where you present conclusions and recommendations.

EXAMPLE: PERSUASIVE REPORT

Like the other two types, we can divide this into steps. The introduction defines the subject and presents your qualifications. What is the need? What are you trying to address? Many project issues are kept alive because of people's ineptitude in defining the issue or situation. Avoid anything here about the approach or solution. What will happen if the need is not handled? How would the project or organization be if the need or situation were addressed? You are now presenting how nice the world would be if the situation was addressed. This is where you get commitment from the audience. They are now ready for your solution. Don't spend time on the details of the solution; spend time on how the solution produces the result that you presented above. The close—explain the specific steps that the receivers must follow. "Pick up that phone and call . . . " "Fill this out," are examples.

PRESENTING WITH THE PROJECT TEAM

In a project the situation can arise where you have to work with the project team to develop a document or presentation. Writing with a team or committee is almost impossible. There are many different interests. People have their normal tasks and do not assign a high priority to the work. There are a few pluses to a team approach. Collectively, the project team brings the following to the table:

- Variety of skills
- Breadth of knowledge
- Memory of the project and details
- Multiple, differing views and perspectives

Clearly, there is an abundance of information and expertise to draw upon, if you do it right. Get the team or group talking about the project and the subject of the presentation. You are attempting to establish agreement on basic facts. You are establishing a common memory. As you do this, you must establish consensus. After your meeting(s), you will have an understanding of the situation and a definition of the message you seek to convey. Next you have to develop an outline for your team presentation. A group cannot easily come up with an outline. You should develop a straw man outline and present it to them. You do the same with the tactical decisions we have presented.

MAKING INFORMAL PRESENTATIONS

Typical examples here are project status presentations and presentations regarding issues. What does the audience want to hear? What we said in an earlier chapter—status and then issues. Simply put, they want to know if the project is on schedule and within budget. If there is a problem, what is it? What actions are being taken? What is the schedule? Here is a presentation order we have used:

1. Identify the project.
2. Bring the people into the project by summarizing the status of the project at the last meeting.
3. Give the current status.
4. What is currently going on?
5. What are issues, problems, and opportunities and what is being done?
6. Answer questions.

Sounds direct, eh? It is, but you would be surprised at how many people plunge into discussing the problems. The audience gets lost and the presentation has to be repeated.

Another type of presentation is that for addressing an issue. Here you would start with identifying the issue. Explain the importance of the issue. Give your order of presentation. Now give highlights of what has been done in the past to address the problem and why it failed. Briefly summarize your approach. With this background you can now summarize the findings. Do not defend your findings. Instead, explain the impact of the findings.

Here are some general presentation tips:

• Look at the audience at all times.
• When you get a question, even if it is loud, repeat it. Repeating a question allows you to get at an issue.
• Answer the question directly. Don't beat around the bush.
• What if you do not know the answer? Say so and that you will find out the answer and give it to them.

VISUAL AIDS

In preparing a presentation you would first make the strategic decisions discussed earlier. You would decide on order (through the outline). You would start to assemble evidence. These are the visual aids. Here are some additional suggestions for preparation:

- Minimize the slides or overheads with text. Otherwise, the benefits of the visual graphs will be reduced.
- When you show a chart, carefully explain the axes and what the graph shows. After describing it, explain what it means and why it is different.
- Do not overdo it with multiple fonts and colors.
- Make sure that the words on charts and figures are clear and cannot be misconstrued.

For GANTT charts you might start with a summary chart of 30–40 milestones and indicate where the project is in time. Then the next GANTT chart zooms in on the tasks for the specific period. This gives continuity. Otherwise, if you show the detail first, the people are lost. After showing the detail, you should now pull out a general GANTT chart that shows the future. The sequence was general-present, detailed-present and general-future. Avoid PERT charts in general. It is very difficult to follow the small lines between the boxes. What if someone asks a question that is answered in a latter slide? Do you answer it now? Do you tell them that it will be answered later? Here is what we propose:

1. Answer the question by showing the future slide.
2. Avoid details by indicating that it will be addressed.
3. Return to where you left earlier in the presentation.
4. When you get to the presentation place where you jumped, bring them back and remind them.

You always should answer questions directly. What if people raise detailed questions and you have only a general presentation? You should develop several additional overheads or slides that reflect detail on issues that you think they will ask. You show what you think if you successfully anticipate their questions.

How should you involve members of the project team? You might want to give them visibility by having them give a presentation. Or, you might want to bring a different member of the project team each time to focus on a specific aspect of the project. Follow our suggestions in considering a web page. Check out what other companies and people have done with web technology. In particular, consider some government and public sites that publish project information on the Web.

EXERCISES

The place to start is to review some of the presentations for your project or past projects. Put yourself in the position of a manager in the audience.

Do the presentations follow the outlines we have developed here? See if you can write down the message conveyed—not the message stated.

Now try out some changes yourself. To limit risk and exposure you should test these out on the project team. This should be a sufficiently diverse and critical audience. For management, test it out on progress or status reports first. Keep in mind that there are many benefits if you become more adept at presentations and preparing documents. First, and most importantly, you are getting the message across. Second, you build your credibility. Third, you can cut down on the time to prepare the materials.

EXAMPLE

The guidelines presented in this chapter were followed for each medium employed. The following mediums were used:

- Electronic mail—progress reports, notices of meetings, and providing information.
- Internet—access to databases related to reengineering and sharing of reengineering methods.
- In-person meetings—very frequent.
- Groupware—used extensively for tracking issues.
- On-line help and tutorial systems—used for a Windows-based help system for the project management application software.
- Pagers and beepers—infrequently used because these were sometimes viewed as irritating by the staff.
- Voice mail—used for brief updates and messages; no extensive messages were left.

A standard formal presentation on the project was maintained and updated regularly so that presentations could be done on short notice by the project manager.

SUMMARY

The techniques we have discussed are all common sense. The problem is that people who work in projects get caught up in the work and process and forget or ignore the importance of presentations. More times than you think, you are selling. Because project information is often boring and dull and because projects are often technical and complex, you face a double-barreled burden of getting your message across and keeping people awake.

15 Project Management Software

WHAT IS PROJECT MANAGEMENT SOFTWARE?

Project management software is a set of software functions and tools specially targeted at supporting the management of projects. As such, the software attempts to provide planning, tracking, analysis, and output support. Today, project management software is viewed as a key element in our software kit—the differences from the past are features and ability to link to other software and employ the network. Project management software was initially developed in the 1960s and 1970s to run on large computers. In the early 1980s with the emergence of personal computers, several packages for project management on a microcomputer emerged. One was VisiSchedule (developed by the makers of Visicalc, the predecessor to Lotus 1-2-3 and other spreadsheets). The earlier project management software tools focused on scheduling and producing output. Analysis tools and aids were minimal. This observation shows that a primary use of such software is for output to the project team and to management. After all, what good is the tool if you cannot use it for analysis or output? More recent versions of such software have added more help functions, outlining, more examples, a wider variety of output, an ability to handle large projects, increased accounting capabilities, and the ability to handle multiple projects. People also use spreadsheets to track costs and schedules, but this is not project management software.

Software for project management is different from other software such as word processing or spreadsheets. First, it is used less often than other categories of software. Second, few people use it—most see the output from the software. Third, it is possible to customize your use of the software far more than many other types of software. Fourth, it tends to be more expensive than some of the common types of software. Finally, because such software is not as popular, there are fewer seminars, classes, and books available. This brings us to our first recommendation—if you use project management software, pick a popular and known software package. That way, you will not have to explain and defend your selection.

You can spend from less than $100 to over $5000 for project management software. It depends on what you want to do and how much time you have to spend on it. What you get out of project management is basically tracking and graphical output with some analysis support. On the other hand, what you put into it includes the initial learning curve, initial use and setup of the software, time required to embed the software into your everyday or every week schedule, and the effort required to become proficient. We assume that you, like us, have only limited time and effort available to learn and use the software. In this chapter, we want to focus on evaluation, selection, and use of the software from this realistic view.

Table 15.1 presents a series of features and capabilities of microcomputer-based project management software. With your own use in mind, you can later rate these in one of three areas: those that are important; those that are nice to have and why; and those that may not give you much benefit. Clearly, a project management system should be able to input resources, tasks, dependencies, and milestones. It should then produce a schedule and other reports. Now consider some of the features in Table 15.1.

- Multiple projects—The software should be able to link projects together to get an overall project picture.
- Graphs and reports—The system should support a variety of graphs and reports (including PERT and GANTT) in different formats with different printers.
- Filtering—Filtering is the capability to extract a set of tasks and milestones from a project for analysis and reporting purposes. Filtering can be based on specific resources, time periods, dependencies, and other factors. Filtering gives you some of the capability of a database—flexibility.

TABLE 15.1   Features and Capabilities of Project Management Software

Handle costs in terms of standard, overtime, and per use costs per task.
View several projects at one time.
Relate one project as a subproject of another project.
Relate several tasks or milestones in one project to another.
Allow several projects to share a common pool of resources.
Be able to compare planned versus scheduled versus actual schedules.
Allow for multiple calendars for resources and projects.
Support filtering based on various criteria to restrict which tasks are considered.
Support a variety of printers and equipment.
Support interfaces into and out of spreadsheet, graphics, database and word processing
    software.
Support flexible reporting.
Provide PERT charts based on different levels of detail.
Support different formats for GANTT charts.
Support user interface and screen customization.
Preview a report on the screen prior to printing.
Support outlining of tasks under general tasks.
Support resource leveling.
Offer manual or automatic updating of schedules (to allow for bulk data entry).

- Multiple user access—shared access to project files is available through the network.
- Export and import—The system should support a range of file formats for importing information and outputting data for use in spreadsheets, graphics, database management systems, or word processing. Importing is useful because it is desirable to enter data into a spreadsheet quickly and then to bring it into the plan (e.g., tasks can be entered in such a way).

All of the software packages listed in Table 15.2 cover the basic features. Software differs not only in capabilities, but also in the environment. Some work only in a specific Apple MacIntosh or PC environment such as IBM or compatible, whereas others are Windows, Windows/NT, or UNIX-based. These factors should be considered along with price in eliminating some of the packages. The differences between inexpensive, moderate, and expensive software often lie in additional features, networking abilities, and more functions such as cost and expense projection, and so forth. Some questions you should ask in software evaluation are listed in Table 15.3. This will be expanded later in the chapter. Note that the level of expertise tends to peak unless the person is willing to expend enormous effort to learn detailed additional features.

Future features of project management software will likely include more Internet access, support for integrating schedules across a network, and

TABLE 15.2   Some Project Management Software Packages

Name of Product	Manufacturer	Environment
Timeline™	Symantic	IBM/compatible
Project™	Microsoft	IBM/compatible
Instaplan™	Instaplan	IBM/compatible
Project Scheduler™	Scitor	IBM/compatible
Mac Project™	Clavis	Apple MacIntosh
Project Management™	Primavera Systems	IBM/compatible
CA-Superproject™	Computer Associates	IBM/compatible
Flow Charting ™ (for graphs)	Patton & Patton	IBM/compatible

improved links with other network based software such as groupware and electronic mail.

WHY USE PROJECT MANAGEMENT SOFTWARE?

People often use project management software to meet some very specific needs. The software must be able to rapidly produce project reports for meetings. It should track and reveal dependencies, and should be able to export information to word processing, database, or spreadsheet software

TABLE 15.3   Software Evaluation Criteria

Compatibility with software you currently use
Ability to operate on a network
Internet linkage
Integration with other software
Features and capabilities that you need
Ability to run efficiently (not minimally) on your computer system
Ease of use and ease of learning
Availability of reference and training books on the software
Ability to handle projects of different sizes and types
Compatibility with your printer
Readability of the software manuals; availability of tutorial
Extent of on-line help for software
Software has had several previous versions to shake out errors and improve performance
Operates on the network environment in your organization
Compatible with software other people are using
Ability to produce reports acceptable to management
Ability to operate within your work flow and business process

so that you can embed information in reports. Why use it? It makes life easier. It is easier to use project management software to produce GANTT and other charts than a graphics package.

Using project management software is also wise politically as long as it is not taken to the extreme. The software is useful for the project team in that it indicates stable use of a tool. It shows structure to management. It also can get you out of arguments on issues such as "what happens if . . . ". The software tool is then the expert analyst.

Project management software does not manage the project for you. It has limitations. The critical path that the software identifies will not often be the real management critical path. Some software is not conducive to tracking and accumulating costs. It can compare schedules and do "what if" analysis if tasks slip. The software is more like a specific spreadsheet or database tool as opposed to a general purpose tool such as word processing.

Like everything else, the project management software can be abused in several ways:

- The software is acquired and never used.
- The software is used as a drawing tool.
- The software is used for timekeeping and budgeting and falls under its own weight.
- The software project is too large and unwieldy so that using the software for the entire project in one piece is not feasible.
- The project manager gets too involved in the software and keeps reorganizing the project and exploring the software—the project suffers.

To avoid abuse, we will be giving you some hints in the set up and use of the software. Our fundamental belief is that if you handle several projects that involve over 100 tasks and several people over a time period of at least three months, then it is probably worthwhile to use the software—providing you realistically plan your own time.

HOW DOES PROJECT MANAGEMENT SOFTWARE WORK?

We know that project management software takes as input the structure of the project and the tasks, resources, costs, and other input and then updates the schedule. Many software packages allow for schedule comparison as well as for "what if" analysis in addition to graphic and tabular output to paper or disk of the schedule. These are the steps in using the software:

1. Set up the basic schedule information: name of project file, name of project, project leader, text; input milestones, tasks, and dependencies between tasks (this is the project template mentioned earlier); input resources and relate these to the tasks.

2. Estimate the effort required for each task. This effort will result in a completed first pass or initial baseline of the schedule.

3. Periodically update the schedule by indicating tasks completed started, delayed, and so forth, as well as changes in resources. This updated schedule can now be used as a basis for project meetings and for comparing it with the baseline.

4. On an as-needed basis, perform what if analysis using the software and data.

As you can see, the effort using this approach is limited. The schedule might be updated every other week or even monthly for periodic meetings. It is certainly not used every day. When, however, we add timekeeping and cost analysis to the scope of use, we have to enter much more data, verify more information, and generate more reports. This is overkill. There is enough to do in the project without generating busy work.

LEVELS OF USE

We can identify at least five distinct types of activities that relate to the use of the software.

REPORTING

In this application, data is input for the purpose of producing graphs and output for meetings. This is often less work than using a graphics software package. This is a computer form of "etch-a-sketch" where the software supports drawing but its scheduling features are not used and resources are not applied.

TRACKING

Here the project work and effort are periodically logged in terms of completed tasks. The system is used to track progress and compare it with the approved plan.

ANALYSIS

Resources are assigned to tasks. The system is used to perform analysis on moving tasks around, changing task dependencies, changing resources and tasks, and then seeing the impact on the schedule. Many packages allow this analysis. The problem is having the time to systematically lay out alternatives and to then input these one at a time to see the results. Otherwise, many changes could be input and the results would be muddied by the cross-impact between changes.

COSTING AND ACCOUNTING

In this mode, costs are assigned to resources. This can be done as a fixed resource or based on how much of the resource time and effort is consumed. Most project management software systems lack flexibility in handling costs as well as interfaces into budgeting and accounting systems.

TIMEKEEPING

The hours and tasks worked on by each member of the project team are entered into the schedule on a regular weekly or biweekly basis. Updating can occur through the network, reducing the data collection effort by a single person. Reports can then be generated on schedule versus actual resources consumed, costs, and so forth. A company already has a payroll system. Putting timekeeping on top of this creates extra work. Moreover, because the input is not used to compute pay and produce checks, the validity and completeness of the information are suspect.

The additional time consumed is not just in working with the software. You have to carefully design what and how to use the software for these additional tasks. Remember, "The more you want of the software, the more it wants of you."

─────── GETTING STARTED WITH THE SOFTWARE

Using the software is discussed before evaluation because your evaluation skills and knowledge will be improved if you know the situation facing you. Obviously, the first step in getting started is to open and load the software. We recommend that you install all options of the software right away to avoid problems later. Now for the data. The data files should be stored in a separate directory that is backed up. The data files do not need to be in a subdirectory on the same directory as the software. You should probably keep the data away from the software anyhow. That will reduce the amount of virus checking and data compression you will be doing.

After you have loaded the software, you now need to set it up for operation. Some of the process is similar to that for word processing; other items are unique to project management. Some of the setup tasks are

- Defining working and default directories;
- Identifying printers—naming the two most common in your office in case one breaks;
- Defining colors for the screen and other appearance options.

Another option is the working calendar. This alows you to enter the holidays and days off. Often, the software allows this to be applied to individual resources. For each day you can enter the start time and end time. In general, you should maintain a standard schedule. Otherwise, you will have to update and verify several schedules.

There are also options in layout that are important and unique to project management. One is the format of a task. A task description contains X characters; resource lengths are Y characters; and so forth. There are options on what items are visible on the screen. Although each software package is unique, here are some general guidelines:

1. Keep the task description length to no more than 35 characters. This makes the description field long enough to be descriptive and short enough to be readable with other information.

2. Restrict the resource field that shows on the screen to 8 characters. If you use abbreviations for resources and do not assign too many resources to a task, this is sufficient.

3. Granularity—You can specify whether a character on the screen equals an hour, a day, a week, a month, and so on. Because you initially want to see the entire schedule, it is a good idea to select at least a week or month. You can later change it with a keystroke or two.

4. Many project management software systems allow for either project update after each operation or a batch update after you have entered or updated a lot of information. Use the manual batch update. This will make data entry faster because you will not have to wait each time while the entire schedule is updated. Imagine how long you have to wait with 200 tasks and an individual task change in task 48! When you have made all changes or entered all data, you can then command the schedule to be updated.

After making these changes, you will probably have to save the results so that they will be the default when you start the software up again.

LEARNING THE SOFTWARE

After the initial loading of the software, the tendency might be to read the manual. Do not succumb. Start a sample schedule that was included in the software to see how it was set up. Remember you are learning how it should be set up with the software as opposed to learning project management. Use the sample project to learn the keyboard and what actions are triggered by specific keys. Some specific ones you need are (1) help; (2) save the schedule; (3) update the schedule; (4) get the menu; (5) get task detail for editing; (6) delete a task; (7) set and delete dependencies between tasks; (8) insert a task; (9) save the task; (10) print a GANTT chart. After working with the sample schedule, go ahead and establish your own phony schedule. Follow the standard conventions and rules discussed earlier in the book:

1. File name for the schedule should be of the form: XXXXYZZZ where XXXX is the name of the schedule, Y indicates the version (B-baseline, L-latest, etc.), and ZZZ an abbreviated form of date. The Y can be used for B, A (actual), and P (planned). For example, we might use SAMPTA15 for a sample schedule in a test mode with the date of October (A) 15. Enter the name of the project. Using long file names, even when possible, can be confusing when viewed through standard file manager software.

2. Enter at least five tasks (A, B, C, D, E) and one milestone. Use the standard abbreviations you have set up as well as the task numbering discussed earlier. Save this schedule.

3. Now set up dependencies between the tasks: A precedes B; B precedes E; C and D precede E and go in parallel to A and B. D can be started two weeks after C starts. Save this schedule.

4. Enter resources in the standard abbreviated form discussed earlier. Assign the resources to the tasks. Usually this can be done by editing the tasks. Save this schedule. This is your template.

5. Now develop schedules for each task using fixed or as soon as possible (ASAP). Set the milestone at the end as fixed. This is a firm deadline. The tasks then have to lead up to the milestone. When you enter the duration of a task, keep with days or weeks, as opposed to hours or months. When you have completed these steps, save this schedule.

You have now created four versions of the schedule that have individually been saved as separate files. You can produce reports for these tasks and schedules. You should produce GANTT, PERT, and resource charts as well as detailed task charts to verify what you entered. Verification using reports as opposed to trying to read the screen forms is often easier and gives you a different perspective after you have been staring at the screen to input the information. This work is useful for several reasons. First, you learn the software. Second, you have examples that are similar to that of your organization for use as handouts and explanations in meetings. Again, when you work with software you want to accomplish several goals at one time.

PROJECT SETUP

We are now ready to begin to establish some standard templates and files that will be useful in working with the software later. What we will be doing here is creating a standard resource file in which there are resources with no assigned tasks. We will also create a file with a work breakdown structure but with no resources. Then we will create a file with both the work breakdown structure and the task list. You will be using the rules and conventions discussed earlier. Again, we remind you to avoid outlining because the indentation of the tasks consumes space (and you only have 35 or so characters available).

To learn more about the software and to accomplish this more quickly, we suggest that you enter the tasks into a word processor or spreadsheet and then inport the file into the project management software. To do this, enter two tasks into the word processor or spreadsheet and save it in the format specified by the project management software. Then start up the project management software and attempt to import the two task or two resource files. If this does not work, then you will need to read the manuals for the project management software. But in any case, you did not lose any data because you did not enter all of the real tasks or resources. Again, you should always test a function with a very small amount of data so that you do not destroy your work.

There are typically a number of fields or attributes that you can supply for a test (depending on the software). Some of these are as follows:

1. Scheduling: fixed time, as soon as possible (ASAP), as late as possible (ALAP). In general, the only things you want to fix are milestones that are immutable. All tasks should be scheduled on an ASAP basis for flexibility. If, for example, you were to schedule five tasks in a fixed mode and there was a shift in plans, all schedules for these tasks would have to be manually changed.

2. Duration refers to elapsed time of the task.

3. Effort refers to the effort required over the elapsed time. The effort can be more or less than the duration.

4. Status—Typically, this can be finished, future, or in process. You should put in tasks that have been completed. This will provide the team members and management with a better understanding of the software. It will provide continuity.

5. Percent complete—Usually, the software allows you to put in a percent complete for tasks that are in process.

6. Resources can be organizations, equipment, software, or any entity, and can be identified as follows:

   a. Short name—This will appear on the GANTT charts and so should be very short to allow for fit. We recommend no more than four characters.

   b. Full name—This can be very descriptive, but must not be ambiguous. Exact organization names should be used.

   c. Calendar for individual resources—Many people run afoul of this. They attempt to be fancy and create separate schedules for each resource. Then when the schedule pushes out tasks in odd ways, you don't how it happened. In general, you should not fall into this trap. Use a standard calendar for all resources.

d. Leveling—Resource leveling is the option that allows the software to change the schedule to level the resource usage across the project. Because this often has a substantial impact on the schedule, it should initially be avoided. If you invoke resource leveling and later wish to change the schedule, you often have to edit each resource to take off resource leveling. Let's see what resource leveling does to two tasks (A and B) assigned to the same person for the same period of one week. If these have the same priority, it may put the first one first and the second one in second place. Or, it may have the person work half-time on each. It depends on how you have set the software to do the leveling. If one task has a higher priority, it will go first. Resource leveling moves tasks that have overcommitted resources.

7. Text fields. These are user definable fields. Candidates might be person responsible for the task, risk level of the task, issues (indicate the number of the issues), action items, and lessons learned.

Resources should include more than organizations and roles such as management and project manager. They should include equipment, facilities, and other items needed to perform the tasks. However, you should only include those resources on an exception basis. Politically, you will use this to highlight the need for these additional resources.

Two specific artificial resources are "critical" and "milestone". Tasks that you know are important are critical resources. All milestones are the milestone resources. Because almost all software packages allow the assignment of multiple resources to a task, this should present no problem. The ordinary resources can be assigned as well. Using the filter based on resources, you can now generate reports based on the resources: critical or milestone. Critical is for the management critical path discussed earlier.

After you have entered all of the tasks or resources and have saved the files, you should produce a set of standard reports to use in various meetings and to explain to people how you set up projects. Again, you are addressing several needs through the same work.

There is one more thing you need to be able to do—filter a schedule. That is, for example, have the software suppress all tasks except those that use a specific resource. You should do this for not only learning it, but also because you will need to be able to show someone their tasks in the project.

Once you have the structure of the project, you can then select the specific tasks and enter dependencies. This will be saved as a new file so you will not disturb your general file. Attach resources to tasks and save the results. Make sure that you input a minimum number of dependencies. Dependencies should be major and occur logically and should not be based on the same resources or some other reason related to how things are to be done (as opposed to what is to be done). Save the schedule and produce reports. You will probably make mistakes that can be detected from the PERT chart for dependencies and the GANTT chart for resource assignment.

You now have a schedule without specific dates and times. Enter the latest schedule status for the schedule to date and save this as a separate file. Update this file two weeks later and save it separately. Use the software to compare the two schedules.

PRESENTING THE SCHEDULE TO OTHERS

If this is the first time people in your project have seen this project management software output, you need to get them used to the appearance, structure, and some basic capabilities of the software. You also need to tell them what its role will be and how it will be used. How do you do this in an interesting way? Here's a method we have used with some success. Invite the people in the project one at a time to your office. Show them the reports and hand them a copy. Then turn around and show them how you can update the schedule and compare two schedules. You might want to show them how to filter a schedule. You are not training them—you are showing them that the capability exists and that you are on top of the software. This should not take more than 20–30 minutes. If it exceeds this, you are trying to show off or you are training them. During this period, you also must explain how the software will be used. Once you have established the baseline schedule, you can disseminate it through the Web and server so that people can review it, and with proper security, apply updates.

USING THE SOFTWARE OUTPUT FOR PRESENTATIONS AND MEETINGS

After reading this title, you might think that we are going to talk about fancy fonts and color output. You would be wrong. We want you to use black and white output that can be copied and 8-1/2 × 11-inch paper. We suggest staying away from pretentious presentations that detract from the analysis and comprehension of the project information. Our first rule in the use of the output is to have a minimum level of output and then only expand this on a very selective basis. There are several approaches:

Level 1

Project is basically on track; you are giving out routine information on the project. Use only the GANTT chart. Show only recent major milestones and current and near-term future tasks. Print out at most 100 items.

Level 2

There is a project issue at hand. It may relate to a specific resource or set of tasks. You should produce the GANTT chart along with selected detailed reports and charts. An example might be to filter on all tasks and milestones that are dependent on the late tasks. Another situation might involve a critical resource or department. A filtered project can be established using the specific resource. Another report that is useful is the task detail report, which contains all of the information on a specific task.

Level 3

The project requires substantial change. You have to explain the situation and propose a course of action. To explain the current situation you should develop a projected schedule that illustrates the effects of slippage and other problem areas. You should avoid too much detail. Otherwise, there is a danger of being involved in finite analysis of the problem and not the solution. After presenting the problem, present a work breakdown structure that is your solution in restructuring. Get an understanding of how this structure is different. Then jump to the new schedule with the changes. Explain the new schedule by comparing it to the old schedule.

Level 4

The project is in major trouble. Either a major change or termination of the project should be considered. You should always consider shutting down the project. We present three scenarios: (1) do nothing different and watch the resources go up in smoke; (2) invoke major change and restructuring; (3) kill and phase out the project. The first two schedules can be generated in a way similar to that of the previous level. For the killing of the project, you need to do more work. In addition to a new task plan, you need to develop a political explanation acceptable to the organization. You also need to develop a rapid decline scenario that is still orderly. Address what can be salvaged from the project and what can be learned from the project. We will discuss these items in more detail in a later chapter.

USING THE REPORTS FOR COMPETITIVE ADVANTAGE

A project management software package typically has a range of reports available. We have identified some of these already. A more complete list is as follows:

- GANTT chart
- PERT chart
- Task detail
- Resource histogram—this shows the usage of a specific resource over time.
- Cross tabs—tasks are rows and resources are columns. This shows which resources are required for the tasks.
- Status—the status of currently active tasks

But this is only part of the picture. You also can select the subset of the plan to be reported. This is where you can gain advantage. Be selective filtering on tasks you can produce reports and charts that help your argument and make your point. Here are some examples:

- A department complains that they are too involved in the project and that it is consuming too much time. You should filter by each department and produce histograms for each department. This should illustrate fairness or equity. You should also produce the cross tabs mentioned above.
- A department or manager wishes to delay tasks. You can filter on all tasks dependent on the tasks that they propose to delay. This will show the impacts of delay.
- Some people are not working on their tasks to the extent that is required. The same filtering approach can be used.
- People are confused by the project because it is large. You should focus on the Critical tasks (those labeled with the resource Critical).

UPDATING THE SCHEDULE

Now let's suppose that the project has to be updated. We have talked generally about this. It is time to get precise. Obviously, you are not going to call everyone on the team every day and then post the percentage complete to all active tasks. You will probably contact every one once a week and obtain a status report. Tasks that are completed can be marked at 100% complete. New tasks that are started can be marked as started with the start date changed, if necessary. We recommend that you do not change each active task that is still in process unless there is a change or slippage. In collaborative scheduling where each person updates, managers must review the schedule. When updating the schedule, tasks are marked as complete and dates are entered for actual start and finish. The baseline schedule which was approved before is not changed.

VERSIONS OF THE SCHEDULE TO SAVE

After reading the preceding section, it is clear that you will be generating a number of schedules. In addition to the baseline schedule, there are updates to the schedule. There is the last updated version. Filtered schedules are usually not saved separately, but instead are generated as needed. Furthermore, there are schedule projections for specific scenarios as outlined above. In the file naming convention discussed previously, the Y or middle element can be used to include planning, actual baseline, and so forth. The project name should explain the purpose of the anlaysis and what was done. Why should we save these versions? We may need to review these if management asks why certain decisions were taken. Moreover, if you save the reports in paper form, you will lose flexibility of analysis because there is much more information behind the report that has not been printed.

SOME HINTS—POLITICAL AND OTHERWISE

When reports are printed, many often include the date and time that the report was printed. It can be embarrassing if you did the report at 5:00 A.M. and it is noted by people in a meeting (because the time is printed at the top of the report). Here's a suggestion. Prior to printing the reports, exit the software to the operating system. Change the system date and time to a more politically acceptable day and time, say two days ago at 6:00 P.M. In DOS this is done through the DATE and TIME commands. Similar commands are available for Windows/NT and other operating systems. After changing these, restart the project management software and print reports. Remember to change it back after you have finished generating the reports.

Do not attempt to be an expert in the software or attempt to show off your software skills. It may make an impression exactly opposite to the one you want. If you attempt to explain nuances and details about the software, people may think you are a computer nerd. You will be viewed as a techie as opposed to a manager.

Be involved. If you depend on a staff member to prepare the reports and do the analysis, you will lose credibility because you cannot answer questions. People will figure out who to ask and will bypass you when they have questions. What should you do? At a minimum we feel you should acquire some expertise in using the software. You should have been involved in the set up of the software. You should also be involved in data analysis. What can you avoid? Data entry, running reports, and some of the data collection.

Use the software on a regular basis. If you get rusty, you will have to relearn the software. This can be very time consuming and frustrating. Do some analysis using filtering and comparisons of schedules. Filter by milestones and see what is coming up.

Another hint is to know when to avoid using the software at all. We discuss manual methods later, but we can comment that if a project has little risk, is short in duration (less than three months), and does not involve many tasks or resources, it would be overkill to use the software except for practice and learning.

Do not take the schedule results for granted. Too often people become overly dependent on the schedule and software. The schedule is only as good as the input. If some tasks were put in as late as possible (ALAP), then the schedule will likely be delayed. This suggests that you need to audit the tasks for accuracy. You can do this by printing out the report for task detail. This is tedious work, but it has to be done. It is, in fact, similar to auditing the formulae in a spreadsheet.

STANDARDIZATION

A common question we are asked is whether an organization should standardize on a specific project management software package. In principle this appears to be a good idea. For shared information common software is essential. Unfortunately, real life experience indicates that some software is better suited for small to medium projects; other software is better for large projects. Another thing to consider is the set of responsibilities that are incurred with the software endorsement. You may then have to provide training. To avoid this, name a technical expert for the software. Additional manuals and materials may be needed. If you are endorsing the software, you should also identify naming conventions and some of the rules we have discussed in this chapter.

You do have to do something, otherwise, you may have Redundant and incompatible software. A suggested middle ground is to endorse a specific, popular software package that runs on the most common type of PC. Develop guidelines as to when it should and should not be used. Concentrate on middle-sized projects. Large projects have a life (and death!) of their own. Small projects are too much effort. Provide guidelines on the conventions and rules for the software. Have people try to learn as much on their own as possible. You need to encourage self-sufficiency. If there are sufficient users, then encourage them to get together now and then and share experiences.

EXERCISES

Find one or more articles on microcomputer-based project management software system. Sources are *PC Week*, *PC Magazine*, *PC World*, and over 100

Example **269**

other magazines. Typically, a magazine will provide an updated review every year or two. Also, use the Web to search for project management software articles, products, and reviews.

Scan the article(s) by focusing on the comparison tables. Read through the criteria of comparison. If the article selects one system as preferred, analyze why they prefer this particular one. If the selection is based on price, you can downgrade the review because the price on the street is often less than the suggested retail price. Now accepting the review approach determine which is the easiest to use and which can handle schedules of 200 tasks or less the best. Also, consider which has the widest range of output and supports filtering.

You will see by performing this analysis that the same package is not preferred for all criteria. By reading several articles written over several years you will see the changes in criteria that occur due to improvements in hardware and software. In general, you will be wanting to select software from an established software firm that has provided several previous versions.

Next, go to a book store that offers computer books. Look for manuals on project management software packages. If there are many books on a specific software product, you can possibly infer that the software is popular, but also may be complex. If there are no books on the software, then you can assume that the software has not achieved a level of popularity and installed base to justify a book.

Without spending any money, look around the office and talk to your friends to find out if they have a project management software package. If they recommend one, there are several things you can do. First, you can attempt to use the software to test how it works. Even if the software is several years old, you will now gain exposure and limited experience. Second, find out how the system is used or not used. Often you will find that it sits on the shelf. When you ask why it is not used, you may be met with embarrassment because the people feel bad about spending the money and not using it. Do not ask why they do not use it. Ask questions like who attempted to use it; what types of projects was it used on; was there any training?

EXAMPLE

The project management software used in the project was capable of running on both MacIntosh and PC platforms. There was, in addition, a database that linked to the project management software. The database contained the project information while the project management software served as a front and back end to the database.

Some observations from the use of the project management software are interesting. First, the use of export and import of data to and from spreadsheet software was substantial. A popular approach was to produce a table of resources (rows) versus resource use over time (columns). This could then be employed as the basis for calculating cost-related tables such as earned value. A second observation was that although there was a requirement for a minimum set of conventions on the structure of projects, there was a great deal of flexibility in how individual managers worked with the project data that was downloaded from the project database. This data was read-only and so the main data was protected. Updating the schedules and database was controlled.

Each area within a division, such as final assembly and integration, had standard task names and structure at the area and process level below the area. However, each area manager could customize the tasks below the level of a process. The database tracked at the level of the process. This approach supported both standardization for management as well as flexibility to accommodate different types of manufactured products.

There were four different schedules for each type of manufactured product. One was the schedule for the customer; the second was for marketing. The third was for operations management. The fourth was for the area manager. Multiple schedules allowed for analysis, data control, security, and yet ensured consistency. By the way, these replaced over ten existing schedules.

SUMMARY

Project management software can be very valuable for both management and political reasons. However, it requires substantial dedication of time and focus to successfully implement and use the software. If you think that you will not have the time or inclination to devote to the tasks we have identified, then don't bother. You have to develop the pattern of using the software as a familiar tool, not just something on the side. But remember, you will have to come up with reports and analysis some way.

16 Project Administration

WHAT DO WE MEAN BY MANUAL?

For very small projects of short duration, it is absurd to consider a series of automated methods. The overhead and hassle of using software and other tools outweigh the benefits. You could manually summarize and find the critical path in projects of up to, say, 40 tasks. However, for projects of any size there are still other administrative activities. Here we consider the project file, handling project documents, project meeting notes, and manual methods for tracking and managing small projects. These are critical in small, manual projects because they often consist of the body of methods and tools. We also consider at what point automated methods should be introduced into a manual project.

On the surface these topics must appear bland and somewhat boring or routine. Yet, as we have noted in several examples, they are required to support organization and control.

ORGANIZATION—YESTERDAY VERSUS TODAY

To see what things were like over fifty years ago, you should go to a local museum and see how an office was established. You also can get some ideas from a 1930s movie. Why is this of interest? Because it shows you how people managed work and projects without copiers, fax machines, and computers. If you want to get even more manual, go back to the offices of the mid-1800s in Europe—without telephones and telegraphs.

The first thing that you will observe is that the office of yesterday was compartmentalized. Because labor was less expensive and because there were few aids except typewriters, people had to have very specific duties. Job descriptions were much more precise. There was the idea of apprenticeship where one learned on the job from more senior employees. During the initial period of employment, a person not only learned how to perform the tasks, but also was given clearly identified work standards. Because of the time-consuming manual processes, there was no time to redo everything several times. Moreover, waste was tightly controlled. Many organizations controlled pens, paper, carbon paper, and pencils. People had to take more care in recording information because errors were more difficult to identify and correct.

A substantial amount of training and transactions was supported by oral procedures and shortcuts. With the telegraph many successful companies developed a series of confidential short codes. This not only protected business messages, but also reduced the cost of telegraph service.

Projects were controlled through the use of written records for all resources. Paperwork and accounting records existed for goods, services, and labor. Project status was handled through in-person reports and reporting in handwritten or typed status reports. Issues and questions on projects were handled the same way. This can be seen in the construction of Roman roads, aqueducts, and large structures. It can be seen in the construction of railroads across the United States in the 1800s. An example might be the excavation of a hill for the Union Pacific Railroad. The manager at the railhead would prepare a statement of the problem (removal of earth for the track) and specify the requirements for dynamite, powder, and any earth-moving equipment available as well as additional labor. Headquarter's groups assigned a high priority to the requests and status reports to field managers because time meant money.

Although these examples are interesting, they serve to point out that many large projects have been completed with manual methods. The reasons for their success are many, but some of the key elements we have gleaned from experience are as follows:

- Clear organization of the written documents to avoid any ambiguity, because there was no time to ask for clarification.

- Clear definition of people's roles and responsibilities. Thus, when a message was received from someone, the receiver clearly understood the sender's role and power.
- Discretionary authority within limits. Due to time and distance, empowerment of managers was critical. With the large distance and time gap, the British Empire and the Raj system in India succeeded for so long in part due to Britain's effective training and empowerment.
- Standardized structure for reports on status and issues. Standards facilitated organization and supported completeness of information.
- Although it appears in retrospect that decisions were made in a timely manner, this was not often the case. Due to the communications and control limitations, in many cases, problems were not addressed until they became major headaches at the headquarters level. Managers were expected to resolve the problems at the lowest level within the project without additional resources.
- The decision-making process in projects was more deliberate than today. People had to think through the consequences of actions and decisions because once the word went out, it could not easily be revoked. Important decisions dating back to Alexander the Great and the pharaohs were carried by a courier who not only carried the message, but was briefed to interpret the message in light of the situation that was current. After all, the written word was almost always obsolete and of limited value. The importance of the sense of the decision cannot be underestimated. The sense of the decision was embodied in the overall strategy.

These points are still valid today. Automation in some ways has supported sloppier and more careless work because people feel that bad decisions can be undone more quickly. However, properly used automation can provide time to be more careful.

MANUAL ADMINISTRATIVE PROJECT WORK

Within any project today there are manual parts. How data is collected, how it is organized, how it is documented, and how it is used are all examples. Computers are tools in analysis, data management, and communications, but the core of much of a project is still manual, and probably always will be.

Is the manual part of a project automatic and instinctive to most people? No. Like many other things, manual procedures have to be learned and acquired—not automatic. The manual part is not paper and paper handling; it is the interaction with people and analyzing and organizing the information.

TYPES OF PROJECT DOCUMENTS

Let's open up a typical project and see what types of documents are present.

1. Project description and initial project plan. This sets the stage for everything. If you cannot find this quickly, you will have a hard time responding to charges and questions regarding project changes.

2. Project assumptions. This should be part of the initial project plan. If it is missing or never was done, then the seeds of future problems and misunderstandings have been sown.

3. Project budget reports. These could be computerized or manual budget reports on expenditures and cumulative expenditures versus budget.

4. Project status and meetings. This includes agendas as well as the results of meetings.

5. Project memos. All types of memoranda are written on a project relating to issues, resources, schedule, and so forth.

6. Personnel matters. These are memoranda and documents related to getting project resources, getting rid of project resources, and so on.

7. Project milestones and end products. In addition to physical evidence, there is often paperwork relating to design, requirements, certification, acceptance, and so forth by users, government agencies, and others.

The problem here is how to organize these documents so that we can efficiently find and use them.

The Project File

The project file can be regarded as the repository of the project. In theory it contains all of the items listed above and more. The first organizational step we suggest is that the documents that are technical or relate to milestones be kept in a separate file. This file can be organized by milestone with the start of the file being the first milestone. It is important that this file contains all technical and no political documents. This file is used for many things. First, it can be handed to a new project team member for review. Second, manager can review it. Third, it can be stored out in the open as long as a separate, backup copy is maintained. This corresponds to an architect's drawings and blueprints.

We are now left with the mass of management and politically sensitive materials related to resources, issues, staffing, and other items. A key to understanding a project and dealing with project issues is the chronology of the project. When did things happen? Thus, our next suggestion is to arrange the file chronologically within sections. Because many memos and documents address multiple subjects, it is unreasonable and inaccurate to overly sort the project documents.

Here is a compromise approach that we have used successfully. Separate the management part of the project file into three parts: (1) those dealing with budget items; (2) those dealing with status and meetings; (3) all others. A fourth category applies to larger projects—namely the issues database. The part dealing with budget reports can be sorted in chronological order. Satus and project meeting reports can be stored in the same way. The same applies to the rest. For the issues database, one would first have the most current issues summary at the front of the section. This would be followed by descriptions and analysis of each issue.

There remains a problem—how to retrieve things that are stored chronologically without having to go sequentially through every piece of paper. Our suggestion is to use color-coded tape attached to the edge of the paper. This labels the pages where a specific topic or issue occurs.

MANAGING DOCUMENTS—CURRENT PROJECTS

As a project manager, you receive many pieces of paper at different times. You also receive electronic mail. For electronic mail that is important in the sense that resources or an issue are addressed, you should print it out and store it in the project file. It can be very tedious and time consuming to have to recover notes later. Oliver North in the Reagan White House found this out.

It is very tempting to make notes on what you receive. Typical notes are comments on agreement or disagreement. Do not under any circumstances write on the original document. It will become part of the project file and may fall into hostile hands. What should you do? Keep a separate log by date. This is the project diary—only you know it exists. In your diary you should place an entry for each date even if nothing happens. When a document is received, log it into your diary and record your comments here. If you are going to use a word processor, store the diary on a diskette that can be stored under lock and key. Remember this diary represents your own thoughts. Why keep a diary? It seems like extra work and because no one knows you are doing it, you do not get credit for it. There are several reasons. One is to record your own thoughts privately. A second reason is that you want to learn from the project and become a better project manager. By reading what you thought several months ago, your thinking process about projects will improve.

Now back to the documents you have received. Some documents will require immediate action; many are for informational purposes only. No action is required. Do not file these one at a time. This takes too much time. Keep them in your desk in a to-be-filed category. Once a week take all of these out and sort them. Read through them chronologically. Place color tags on them as needed with a different color for each type (orange— end products; red—issues related, etc.).

The type of project file depends on the project. If the documents are pieces of paper of different size, such as legal and letterhead, then you are almost forced into legal size folders with two holes being punched at the top of the paper. The fastener is at the top of the file. If the documents are all letter size, then you can use a standard three-ring binder.

What should you use for the diary? You need something that is convenient and that has firmly attached pages. This rules out pads. Here are two approaches. One is to use a stenographic pad that has spiral binding at the top. You can write the date from and to on the front along with your name. A second suggestion is to use a hardbound 8-1/2 × 11″ book (similar to ledgers). Whatever method you use, do not use the diary in meetings to record notes. It will clutter up the diary. The diary is for summary information.

MANAGING DOCUMENTS—COMPLETED OR
TERMINATED PROJECTS

From our experience, there are several situations to be considered here. We may open a project file and find chaos. There is no organization. What do we do? The second situation occurs when we are in charge of handling a project that is ending.

As project manager, you have been told that a specific project in the past has some valuable information. You gather information on the project. It is lying all over your desk. Our suggestion is to order the information in chronological order. Then follow our suggestions in earlier chapters on reading the project information from the most recent to the earliest document.

In the second situation, you are going to archive the information. In projects of the 1800s and early 1900s, people followed a logical approach. They created a file for the project and placed it in a cabinet. The cabinet was near the people who might need access on it later. Unfortunately, this is another lost luxury. Because of the cost of office space, the nearby file space is limited. There may be no central files in our building. There is only the black hole of off-site storage. Sold under the ambitious title of records retention, you are told to place the files in standardized boxes and then after preparing paperwork you will send it to archives. Academically, this sounds fine. In theory, you can then request the box. Too bad that the reality often does not match theory. Many records retention methods of medium and large organizations are abysmal. How much archived information is ever found again?

These problems arise in part because the activity is viewed as low priority. Taking a conservative approach we suggest that you place the file in your file cabinet for at least six months. Next, we suggest you test the archival system. Ask the staff or a secretary for a list of things archived. Attempt to retrieve it. What was the process? How long did it take? What shape was the material in when you got it? If the system works, then after six months or a year, you might test it by sending the file to archives.

Before sending it to the archives, make a list of the key documents in the archive. Do not send your diary to the archives. You will be using it for other projects.

TAKING NOTES FROM PROJECT MEETINGS

Over the past 150 years, there have been many methods proposed and used for meetings. In the days before electronics, a secretary might attend the meeting and take notes using shorthand. This precomputer age luxury has largely disappeared. Shorthand has become mostly a lost art.

We have said already that you should be the one taking the notes. You might be tempted to place a tape recorder on the table and record the conversations. Although the recording industry applauds this because it sells more equipment, it raises several issues. First, people are not likely to talk openly if they are being taped. Of course, do not attempt to tape without the audience's knowledge. Second, if you tape it, then you will later have to spend the same time listening to it and making notes.

We want a method that is reasonable for projects of any size and can be used without any technology. Our suggestion is as follows. We assume that you have an agenda that is numbered. Place the number at the top of the page. As the agenda item is discussed, make some small notes. After the agenda item is covered, write the conclusion or resolution in large printed block letters that are easily read. Keep it short and simple. Minimize the use of verbs and excess words. They consume space and time and also may carry a political connotation. You should read to the audience what was covered so that there is validation to what you are writing. Continue doing this through the meeting.

At the end of the meeting, you will then highlight the action items. When you write up the notes of the meeting, there are likely to be few surprises. You also will have minimized the effort.

To test yourself and improve your skills, do this in meetings even if you are not running the meeting and are not the official recorder. Then compare your notes with the official record that is distributed. Your goal is to have your notes in better quality and completeness than the official record. Publish meeting notes on the network and distribute through electronic mail.

INTERVIEW NOTES AND SUGGESTIONS

There have been many books and articles written on this subject. Over the years we have employed the following technique that may be useful.

1. Prior to an interview let the person know what you will be covering and how they should prepare. You should give them a list of 4–5 items (no more) to be covered ahead of time. In preparation, you should minimize what they should do. You do not want them to have to prepare; otherwise, you will likely have the meeting deferred because they do not want to spend the time preparing. But have them bring anyone to the meeting that they wish.

2. It is useful to have two people from the project show up for the interview. Often, this is not possible. However, if it can be done, there are several benefits. First, one person can take notes while the other one asks questions. This is useful because you may lose the train of thought when you make notes. Second, two people can later compare mental notes to clarify any questions. If you are doing this alone, then tell the person that you will be taking notes as you go. When you take notes, you will mention what to do next.

3. In taking notes, follow our suggestions for the project meetings previously discussed.

4. At the end of the interview, agree on a list of action items. One is when you will get back to them with a report on the results of the interview. A second action item is any follow-up on data collection.

5. The entire interview process should be manual. Do not record it. Do it in person—do not use the telephone.

6. Write up the interview results within two days after the interview. Otherwise, you will forget something.

7. When you give the interview notes to the person, tell them that unless you hear from them, you will assume that the interview notes are acceptable. The default, doing nothing, is the path of least resistance and should favor you.

TRACKING ISSUES MANUALLY

Now we turn to the projects that are too small or to situations where there is no software available. You should deal with the following:

- How to identify and track project issues;
- How to track the project itself;
- How to report to management on the project.

Figure 16.1 is an example of a form that you can use for an issue. Figure 16.2 provides an example for the issues overall. Our forms should only be used for guidance. You should adapt these to your own needs. The title of the issue should be worded very clearly and carefully. In roughly 50 characters or less, you have to describe the entire issue and its ramifications. Some people phrase the issue in the form of a question. This is good because it provides focus. It sometimes does not work if the issue is too complex.

```
Project:  _____          Issue ID:  _____

Descrpt:  _____

Type    :  _____          Priority:  _____

- - - - - - - - - - - - - - - - - - - - - - - - - - - - - - - - - - -
Date       Status Code      Assigned to            Status
                            Comment
_____      _____     _____        _____
                            _____
_____      _____     _____        _____
                            _____
_____      _____     _____        _____
                            _____
_____      _____     _____        _____
                            _____
- - - - - - - - - - - - - - - - - - - - - - - - - - - - - - - - - - -
Closed out:  _____

Resolution:  _____
```

FIGURE 16.1 Sample form for an issue.

```
Project:  _____              Date:  _____

- - - - - - - - - - - - - - - - - - - - - - - - - - - - - - - - - - -
Issue ID    Priority   Status       Comments

_____  _____  _____        _____

_____  _____  _____        _____

_____  _____  _____        _____

_____  _____  _____        _____

_____  _____  _____        _____

_____  _____  _____        _____

_____  _____  _____        _____
- - - - - - - - - - - - - - - - - - - - - - - - - - - - - - - - - - -
General Comments:  _____

_____
```

FIGURE 16.2 Example of a form for tracking issues.

Then you may wish to abbreviate and put labels in the title as to the type of issue it is.

We note here that these forms can be implemented electronically in electronic mail (least useful), electronic forms (of more use), on the Web, and groupware (most useful). Electronically in groupware these forms can be distributed for review.

On the form is a place for priority. We suggest a simple manual system of three priority levels. Level 1 has to be addressed immediately. Level 2 has to be addressed during this project phase. Level 3 has to be addressed some time. Your priority system is working if most of the issues are of priority 2. Few issues should be of priority 1 or 3. Why not more of priority 3? Because, priority 3 consists of background issues. Having too many of these may indicate that there are significant latent problems in the project that are not being faced.

There is space in the form for the type of issue. You need to define your own types. You might be able to use the following list as a starting point for classification.

- Technical issue, related to how the work is done
- Managerial, political issue related to resources, milestone review, and so forth
- Structural issue within the project
- Resource issue
- External issue, from another project
- Specific task issue

It is clear that an issue can be of more than one type. However, you should keep all issues tied to one or two types. Having too many types makes the type worthless.

What is the value of the type label? It can show you where the problems are arising. During even small projects over several months, you will likely have a number of issues. As you see that a number of issues have common elements, you start to see that the issues are symptoms of an underlying problem or issue. This is like a leaky roof. You first see some minor water damage on the ceiling in one place. Then you see another. The direction points to a specific section of the roof. This example also serves to indicate that until we see the leaks and think about it, we are unlikely to attempt to fix the roof.

Note that there is additional identification information—the date that the issue surfaced and the number. All issues should be numbered so that they can be referenced in meetings and reports. Use a standard sequential number. We use a system that ends in 0. That is, our first issue is 10; the second is 20. The reason is that one issue may later be decomposed into subissues. This allows for the breakdown of the issue number 10 into 11, 12, 13, 14, and so on. The date of the issue should be the date that the issue was acknowledged. It is not the date that someone thought of it. Dating an issue when you first heard of it leads to the problem that people will ask why this had not surfaced earlier.

The source of the issue is useful because the source can be outside the project. Examples are management, a contractor, a vendor, another project, and so on. Naming the source helps to understand something about the issue. Our suggestion here is that you use a standard list of sources for uniformity.

The description of the issue should define the issue and its dimensions. It should start in the form of a question. Amplification can come in the form of additional remarks to indicate the various dimensions of the issue and why it is important. The description should support your priority and type assignment.

The implications of the issue should include the following:

• What would happen if nothing was done with the issue?
• What would be the benefit if the issue was resolved?

Status of the issue refers to the state in which the issue resides. Our list of states includes (1) not assigned; (2) assigned, but not active; (3) active; (4) resolved; (5) tabled; (6) merged or combined with another issue.

Any issue usually has to be assigned to someone for analysis. Even if you are handling most of the issues, put your initials down on the sheet. Just because the issue is assigned does not make it active. That is why we have two separate status codes.

The comments section of the form refers to the intermediate notes regarding the issue. Each comment should be dated. The length of the comment should be limited to one or two lines at the most.

The section on related issues should identify the issue(s) by number and then indicate the nature of the dependence. It should not go into detail. After all, the form does not take the place of your memory. The resolution of the issue should start with the method of resolution. This is followed by the actions to be taken as a result of the resolution. The date of resolution should be the date that the project team acknowledges that the issue is resolved. In using this form, use a pencil so that you can change status codes. You can, of course, put this into word processing and create a standard form.

Moving to the form for all issues. If you can use a PC-based simple database, then sorting and organizing the issues is much easier. But if this is not possible, you might use the form we have designed. As you can see, we have numbered the issues on the left. The status is on the far right so that you can identify quickly which are active. Next to status is priority. In that way, you can associate status with importance quickly without having to scan across the page. The descriptive information is on the left.

TRACKING THE STATUS OF THE PROJECT

We assume that you have a list of tasks. Because you are doing this manually, you will only have a task number, who it is assigned to, when it is to start and end, and its status.

Figure 16.3 gives an example of a form to support this. Notice there are no dependencies, costs, or detail. These things are all in your head. You have to have a bare bones approach to project management. After all, you are spending more time and effort tracking the issues we have discussed previously.

This list of tasks is really a to-do list. It shows organization and structure. Tasks that have not been assigned are at the bottom. If you are using word processing, then divide the tasks into those that are done (at the top), those that are active or soon to be active (the middle), and those in the future (at the bottom). In that way, when tasks are finished, you can use cut and paste and move the tasks to the top list.

```
Task ID: _____                    Type   : _____
                                       Phase  : _____
                                       Status : _____

Description: _____

Assigned to: _____

Start- Scheduled: _____   Finish-Scheduled: _____

        Actual   : _____          Actual   : _____

-------------------------------------------------------------
Tracking
Date        Status        Comment

_____     _____    _____

_____     _____    _____

_____     _____    _____

_____     _____    _____

-------------------------------------------------------------
Date          Comments

_____    _____

_____    _____

_____    _____
```

FIGURE 16.3   Example of a task form.

REPORTING TO MANAGEMENT ON THE PROJECT

If this is a small project, then you may not be reporting to management often, because the project may not be viewed as high priority. However, you should still provide regular reports and be able to explain issues as in large projects. For project reporting on a manual project, we have used the form in Figure 16.4. It contains the title and description of the project, the project manager, and date. Next it contains the space for budget versus actual expenses. A first thing management might want to know is whether the project is on budget.

The next section addresses whether the project is on schedule. For a small project this can be done by a list of milestones accompanied by the expected date of completion and the actual date of completion. The final section of the form gives a list of active issues and their status and priority. This form is relatively easy to create and update. You may not wish to use

```
                     PROJECT REPORTING FORM

Project: _____        Date: _____

Desc.: _____

Manager: _____

Budget:
                  Project to Date    Period to Date    Total
       Budget     _____    _____   _____

       Actual     _____    _____   _____

       Comments: _____


Schedule-Milestones:
       Milestone                     Planned      Actual

       _____  _____     _____

       _____  _____     _____

       _____  _____     _____

       Comments: _____

Issues:
       Issue                                  Priority   Status

       _____   _____   _____

       _____   _____   _____
```

FIGURE 16.4 Example of a manual management reporting form.

graphs due to the labor involved. Instead, you would use word processing and update it on a periodic basis.

HANDLING MANY SMALL PROJECTS

It is sometimes the case that you have to oversee a series of small projects. The guidelines above address each project. What is left is to have a project summary sheet. An example is in Figure 16.5. This figure contains the descriptive material on each project on the left. The current critical milestone that is being worked on is on the right. You can also assign priorities for each project. In any event the projects should be ordered in either priority or grouped by their interrelationships.

Some guidelines for managing multiple projects manually include the following:

PROJECT SUMMARY SHEET

Project Group: _____   Date: _____

Project Desc.   Milestones   Status

Comments:

Projects   Comments

FIGURE 16.5   Project summary sheet.

Do Not Attempt to Be Fair

There is a tendency to divide your time and attention equally among all of the projects. This is fair, but also stupid. The risk is never the same across all projects. At a specific point in time, certain projects have a very high risk and need attention. Others are proceeding routinely. You need to act like a 35-mm camera with a zoom lens. You need to focus on specific projects at a given time.

Attempt to Achieve Some Economies of Scale of Effort

This pertains to your time and work. When you are analyzing a project, analyze all of the projects at the same time. When you are talking to someone about one project, if they are involved in other projects, ask them about these other projects. This sounds obvious, but there is always the tendency to become overly focused on a specific project.

Perform Self-Analysis across the Projects

Review your role in each of the projects. How do you manage these projects? Even though the projects are different with different priorities, is your role consistent? If it is not, why is this so? Does it relate to the specific people in the project? Is it because of the project itself?

Test Your Memory

We have stressed that you will have to do much of the management and analysis in your head, not in a computer system. To test yourself, sit down and start reviewing each of your projects in your mind. Try to organize these in terms of importance overall. Then try to resort the projects in terms of immediate priority for your time. If you find these steps difficult, you are probably losing touch with the projects.

Where Is Your Time Going?

Measure yourself in terms of where your time is being spent. How much time is consumed in moving between projects either physically or mentally? How have you organized your time in reviewing documents and analyzing issues on various projects? When you are dealing with multiple projects, it is helpful to be able to move quickly between different projects. This is the human equivalent of multitasking in a computer system. Thus, when you review memoranda on a project, take this time and review all memoranda on all of your projects. Try and batch all of the projects together when you are alone. It is just the opposite when you are with staff or managers. In these settings, you want to show focus. If you move between topics too much, you show that you are either confused or have a scattered mind. Not good.

HOW TO ESTABLISH GOOD MANUAL PROCESSES

After discussing the manual work in a project and managing projects manually, we can discuss how to improve the current project management process in an organization. The most successful training and development in this area are done when focus is on the manual process and not on the computer software. When the software is included, the training becomes more mechanistic and focused on the software. To avoid this, the training should focus on the life cycle of a small project.

Effective learning results from people reading and reviewing a good example and then creating their own. In manual systems people learn by doing it for themselves. People also learn from their own mistakes. Thus, one technique is to review a project from each person and give them feedback. People have to be shown that they themselves benefit directly from improved methods. To support learning, a sample small project should be kept for training and reference. A model for what is to be done is very useful.

After people have completed the training, they should be brought back together several months later to show what they have learned. They can share hints and suggestions among themselves. Techniques from drug and alcohol rehabilitation workshops can work here.

EXERCISES

What you should do is clear from the material in this chapter:

- Start immediately keeping a diary.
- Begin to build a project file.
- Set up the forms for manually managing a small project.
- Identify a list of issues and place them in the manual issues database.

In addition, you should review the current practice in your organization. For example, ask yourself how to get your hands on the project file for a completed project. Is the organization open or closed in its approach of sharing information on previous projects? Organizations that do not allow their employees to learn from projects are doomed to repeat mistakes of the past.

EXAMPLE

A major goal in the project (due in part to the small size of the project team) was to ensure that project documents served multiple purposes. For example, if an issue was addressed by the creation of a convention or rule, then the rule would be developed and then referred to in the resolution of the issue (instead of documenting it twice in the resolution and the convention). Documentation for the first project involving final assembly and integration was written in such a manner so as to be usable for other divisions in later projects.

The project file was maintained electronically or in one box (called a banker's box because banking records fit the box). The box could then be hauled to meetings when needed and could be moved easily when a new project was started with a new division.

SUMMARY

What appears to be a very mundane and uninteresting area of project management is very important to the success of the project. Automated projects will still be managed by people. The benefit of manual methods is that they are much more flexible than automated systems and methods. Ideally, a person would learn from manual methods on smaller projects first. Then the automated tools and more advanced methods and tools could be learned and used. In this way, a person could gain perspective on their effective use.

V EVOLUTION, REVOLUTION, AND TERMINATION

17 Project Change and Death

A key part in project management is how we cope with change. Today technology allows for rapid communications and greater availability of information. Issues are often discovered earlier. There is often a desire to get at the issue immediately after it surfaces. Time compression puts pressure on effort required to understand and reflect on the issue. Micromanagement arises more frequently. It must appear as a paradox: we can simultaneously empower people for projects and then turn around and micromanage them. How to understand the need for project change and defining the available options (including killing the project) are the focus of this chapter.

WHAT ARE CHARACTERISTICS OF CHANGE?

Not only problems trigger project changes; projects may change in response to new opportunities, new managers, and new methods and tools. Change also derives from experience. Sometimes in projects there is a need for monumental change to shake things up and alter the status quo. Sources of change include project results (or the lack thereof), the team, other projects, competition, regulation, technology, and management direction. What we can change is wide ranging: project structure, purpose, or scope; leader or team; tools and methods; interfaces; resources and budgets. Next, we can introduce changes suddenly or subtly.

OUR APPROACH FOR IMPLEMENTING CHANGE

As we have seen, a project and its resources respond to change over time. Continuous changes may lead to loss of control. However, we do not want to cast the project in stone because the project then becomes more out of touch with reality. In history it is known that governments change through revolutionary change if they do not evolve.

Our strategy for change is to minimize the number of major changes, and to incorporate many different modifications within a single major change. In this vein we can see that if we make appropriate changes and keep the project stable, we may end up with a better product or system at the end of the project.

TIMING OF CHANGES

Timing is an important issue. The most natural time for project change is at the end of a phase of a project. This assumes that the project has been divided into phases and that there is an opportunity to change or terminate the project at the end of the phase. This presents a chance to update the process and the schedule as well as to change the project.

There are certain conditions when change at the end of a phase is not appropriate. If we want to make fundamental change in the project and alter its direction, then it can be argued that the current project and structure is irrelevant and does not represent the new changed world. Thus, we recommend the following guidelines:

- For major and minor changes with the purpose and general scope of the project unchanged, we would introduce change at the end of a phase.
- For revolutionary change, we would introduce change as soon as possible, regardless of phase, but only after we have done the appropriate planning.

THE SCOPE OF CHANGE

Changing the project schedule appears straightforward. However, when it is combined with changing resources and additional tasks, then we have a larger job. Define the scope first and base it on the analysis of the issues and opportunities that are present. We can then develop a strategy for change and proceed to define alternatives for change. Having selected an alternative, we then move to consider the implementation approach.

THE PRICE OF BEING WRONG

If we are wrong in the changes, we may face substantial and even severe consequences.

Changes Did Not Solve Problems

After change, the project still does not reflect the real world. The changes do not address the real scope of change needed. We have left out some sources of change or have misunderstood them. The effect can be devastating to the project team. Too little, too late.

A leading software vendor was coming out with a new major version of the software. Months went by without the product being completed and so management decided to change the organization of the project. The assumption was that the right people could get the project done. But after the change, things immediately deteriorated further. Morale dropped as did faith in the project management process. Despite even more effort, the project was only partially saved.

Misunderstanding of Changes

The effects of the changes have not been understood. A series of changes have been defined and taken, but now it appears that some of the changes contradict each other. There may be gaps that the changes cannot close. Supporting elements for the changes may have been forgotten. Next, people start to scurry around to cover up the holes and reinterpret the changes—wasted effort and a sign of lack of organization.

Ineffective Change

Nothing appears wrong, but seemingly nothing changed. The same cast is in the play. Was the change a cover-up to just extend the life of the project? Symptoms of issues may have been addressed, but the underlying issues were not.

THE SITUATION

We assume that you have collected a list of issues and opportunities and are considering changes. You are sure that you can't let the project continue as is. Based on what we have discussed, you define the scope of feasible changes. You now think about a change strategy. The change strategy defines what you want to have in place after the changes have been implemented. Table 17.1 identifies a series of examples of strategies. Your strategy may be a combination of these. The scope will establish constraints for the strategy. The strategy targets issues and opportunities. You narrow the field of strategies to a few. Why have these strategies? You do not want to overlook anything. Another reason is that you will need an overall strategy to be used to support the changes when you explain them. Now what? Define alternative sets of changes.

DEFINING ALTERNATIVE CHANGES

With the list, the scope, and potential strategies in place, you can now start defining potential changes. We have included a simple example of the change process in Table 17.2. Each change in itself should address the following:

- How does the change support the strategy?
- Is the change within the scope you defined earlier?
- Does the change address some of the items on the list?

TABLE 17.1   Examples of Change Strategies

A. Change how the team works on the tasks
 1. Structure of project
 2. Use of methods
 3. Use of tools
 4. Potential team additions
B. Change to deal with internal conflict
 1. Team replacement
 2. Potential project manager change
 3. Reassignment of tasks
C. Change to reflect increased importance of project
 1. Modifications to management reporting
 2. Additional team members
 3. Potential project administrator
 4. Expansion of task plan and project plan
D. Change to redirect project toward different goal
 1. Project structure
 2. Changes in methods or tools
 3. Change in team assignments

TABLE 17.2 Example of Change Process

A. External customer for project changes requirements
B. Complete review of project occurs, identifying issues
C. Alternative change strategies are formulated, ranging from incorporating changes into the current project to major changes to project
D. Middle strategy is chosen—project team remains, but is augmented by two new subprojects and additional team members
E. Implementation includes
   1. Creation of two new subprojects
   2. Modification of project plan structure
   3. Changes in management reporting for project manager
   4. Modification of method of project control
   5. Additional tasks and effort for interfaces to subprojects

---

But there are alternative degrees of change that you can undertake. For example, suppose that you are considering changing the project structure. One alternative is to change the whole thing. Another alternative is to change one phase. Still another is to change a particular area of the project handled by one group.

What may help here is to start combining the potential changes. Some individual changes may yield more benefits when combined. Grouping of changes is natural because the impact of the changes in total will be reflected in the revised project plan. For example, if we decided to change the entire project structure and to introduce new staff to the project, we would incur a substantial learning curve, which would slow down the project. So we might trade off structure for additional people.

This effort will help solidify your thinking about possible changes. It will certainly force you to reconsider the scope, list of issues and opportunities, and strategies. It may also help to develop tables.

- Issues and opportunities versus issues and opportunities. The table entry is blank if there is no relationship. If the row is helped by the column, a plus sign appears. If the row is hurt by the column, a minus sign appears in the table. Obviously, you need only fill out the table below or above the main diagonal.
- Scope versus issues and opportunities. The entry is either plus or minus. If the issue or opportunity falls within the scope characteristic, then a plus appears. If it doesn't, then a minus appears. Note that the rows here are alternative views of the scope.
- Scope versus strategies. This table is similar to the one above. If the strategy falls within the scope, then a plus appears. Otherwise, there is a minus entry.
- Issues and opportunities versus strategies. A plus appears in the table entry if the strategy in the column addresses the issue or opportunity in the row.

- Strategies versus changes. A plus appears in the table if the strategy is supported by the change; otherwise, it remains blank unless it conflicts (then it is minus).
- Changes versus changes. This is a table to show the relationship of one change to another (entry is plus or blank). It is minus only if there is conflict. This is how we would group the changes.
- Issues and opportunities versus changes. The entry is plus if the change supports the issue or opportunity; negative if conflict; and blank if no relationship.

Drawing these as tables where the appropriate items are listed in rows or columns forces you to match these up in a formal, structured way. This will be very useful later when you have to sell the changes to management and the project team.

THE REALISM OF CHANGES

How realistic are the identified changes? From the scope we know that they can be done. The issue of realism turns on implementation and impact. How can we carry out these changes within the context of the current project? Here are some specific questions to ask:

- What is the potential reaction to the change within the project?
- What is the possible reaction externally?
- How much rethinking and repositioning is there going to be after the changes are announced?
- What is the effect on management control of the project?

Try to build some vision of what the environment around the project will be like after the changes have been announced.

IMPLEMENTATION STRATEGY

Elements of an implementation strategy include the following:

1. How will changes be grouped in terms of implementation and timing? It is not generally true that all changes should be announced and implemented concurrently.

2. What support structure is needed for the changes? Any change requires some support—management backing, training, procedures, systems changes, and so forth.

3. How will the changes be followed up? What specific steps will be taken after the changes are announced? How will the transition period work?

4. How will the changes be presented and sold to management and the staff?

Let's consider some examples of strategies.

Strategy of Unannounced Changes

In this strategy, changes are implemented without announcement. They are just introduced with no fanfare or acknowledgment. Like the Greek Trojan horse, when enough changes have been made or the situation is ripe, the changes can be announced. This is a favorite approach because it draws little attention and is low key. It lowers anxiety and appears as a natural part of the project process. It is a useful approach when a project needs more gentle redirection. It also provides greater flexibility because very little is nailed down on paper.

The French Revolution Strategy

The French Revolution strategy relies on surprise and on total change. We have used this strategy when a project had to be turned around to avert failure. It requires a great deal of stealth and planning. All of the changes have to be planned and carefully orchestrated. Then the changes are implemented as soon as they are announced. All project team members perform new tasks to adjust to the changes. No one continues doing the same thing. What happens is that a number of people will be forced out of the project. Usually we recommend at least a 30–40% turnover. The project manager and some of the senior project members are replaced. As in preparing for a revolution or a coup d'etat, there must be a new project plan and manifesto. These will be rolled out as part of implementation.

When replacing project members during the revolution, the blame syndrome needs to be avoided. Because you are making changes and the reason for change is evident, there is no need to place blame on people. Be magnanimous in victory. If you place blame and shame on the people you have just axed from the project, you will have nothing but fear on your hands. The remaining people will think that this is what may happen to them.

Why is such strong action necessary? Normally, you can be patient and wait for a poor project manager or team to self-destruct or turn itself around. But the group and team of the project can sustain and feed the project team. It outlives its usefulness.

Grouping Changes to Support the Strategy

Now the question is how to group the changes in a coherent manner in order to get ready to present them to management. Some alternatives are as follows.

Grouping by Specific Issue or Opportunity

This approach demonstrates that the changes address the issues. However, it may be perceived as reactive or defensive.

Grouping by Strategy

This method defines the strategy and shows how the changes implement the strategy. That the changes address the issues is shown by indicating how the strategy addresses the issues. This is more proactive and has the semblance of being more structured and organized.

Grouping by Process, Plan, and Method or Tool

This is a classical approach wherein the process changes for project management are clustered; changes to the plan are likewise grouped as are changes to tools and methods. This approach has merit in that it is at least organized. The problem comes in timing and cross-impact among changes.

Grouping by Time and Dependence

This approach is to aggregate all of the changes in terms of time frames. Changes in later time periods are not announced until later and they depend on earlier changes. The premise of this approach is that many changes that appear to be independent of each other are in fact dependent. The problem with this approach is that it appears tactical in appearance. Here is an example:

A. Stage 1
  1. Change of project team
  2. Changes to part of the project plan reflecting team changes
  3. Introduction of change to methods—not tools
B. Stage 2
  1. More general project structure change
  2. Implementation of tool changes to support the changes to methods

As you can see, we start the change process in stage 1. We don't want to make the overall schedule change because we have introduced new staff to the project and changed the methods. Because the tools support the methods, they are implemented later in stage 2. This stagewise approach supports a learning curve and allows people a chance to buy into the process.

SUPPORT STRUCTURE FOR CHANGE

We have already noted that tools support methods and so act as support for the methods. This is just the start. Here is a list of some of the items to consider in terms of support:

A. Change of project team
  1. Learning from people who are leaving the project

2. Indoctrinating new people into the project
3. Defining new roles and responsibilities for the newcomers
4. Redefining roles and responsibilities for those still in the project
5. Establishment of reporting relationships among the team members

B. Change of method (similar questions for tools apply)
   1. Conditions under which the new method will be effective
   2. Methods of training and exposure to the method
   3. Guidelines for using the method
   4. Where members should go with questions
   5. Documentation of the method
   6. How the method relates to the structure of the project
   7. How the method relates to existing tools
   8. How the method relates to new tools
   9. Expectations and expected benefits

C. Change of project structure
   1. How current tasks map into the new project plan
   2. What happens to tasks that fall between the cracks
   3. What people should do with their current activities
   4. The overall impact of the schedule changes on the actual work

D. Change in scope of project
   1. Method of redirection
   2. Alterations in the structure of the project
   3. Changes in project team structure

FOLLOWING UP ON CHANGES

Wait! If you think you are ready to sell the changes, you are wrong. Sit down and think about follow-up after the changes have been introduced. Here are some of the things you need to consider:

- Have behavior and attitude been impacted?
- Is anything different?
- Have the changes just been layered on top of the old project?
- How do you know that the changes have been implemented?

These questions point to the need for controls, monitoring, and reinforcement as necessary. In over 80% of the projects we have managed, it has been necessary to have substantial follow-up work to ensure that the changes are effective. Bear in mind that the longer you wait for this review, the more screwed up the project can become. We suggest that you start follow-up one week after the change. Repeat each week to determine the changes. You will have to ensure that the changes actually were implemented and that the desired results were obtained.

SELLING THE CHANGES TO MANAGEMENT

Does management know about the problems in the project? If not, the first thing you need to do is prepare them for the changes. You first have to raise the level of awareness of the situation. Related to this is explaining why this is occurring at the current time. You have to ensure that there is no panic, but that change is natural and it is a good time for change.

How do you know if you have succeeded in getting management support? If they want to get too involved, then you know that they lack confidence in you. If they want details, they do not trust you or they may feel insecure about their own positions.

In any event, we now assume that management is aware of the problem. How do you proceed to get their support? We propose a zoom-lens camera approach. First, summarize the problems and issues. For credibility, zoom in on several to show their impact on the project. Second, pull back out and state what is happening overall with the project. Now you have their attention.

Second, present an overall strategy. What you have to do is to build an image of how the problems will be rectified if the strategy is successfully carried out. This gets their attention and, hopefully, their understanding.

A third step is how to show the staging of the changes over time. You show a timeline chart and a list of changes with each point. Explain the changes top down. That is, focus on the impact that you expect after all of the changes in this stage are implemented. Relate this to the strategy. You now proceed to later stages. As you go through this, you also address the measurement and control aspects. This can be followed up by selected detail. Select three changes of different types and go into them in detail. This also helps build confidence in your plan.

Finally, if you get the opportunity, you can show the tables discussed earlier. That is, you can relate the changes to the issues and opportunities. This can also be done through handouts.

What should you expect from management? You seek concurrence. At worst, you expect a fight as some people want to keep the project going. If you think you are going to be in for a fight, then you need to presell the approach with some of the managers in advance. They can later help diffuse criticism.

SELLING THE CHANGES TO THE PROJECT TEAM

You have obtained management approval. So just go out there and list the changes to the team, right? Wrong. Just as you needed an overall strategy and an implementation strategy, you need a presentation strategy. A down-to-earth presentation begins with an acknowledgment that there are problems. There is no effort to list or elaborate. Now cut to the chase.

Go right to the changes. This will put people at ease because you are relieving anxiety and addressing uncertainty. Now you step back and indicate that these seemingly unrelated changes are part of the overall strategy. Describe the strategy. Relate the strategy to the changes. Now you should return to the changes and discuss their implementation and what is expected of people.

This approach (1) reinforces the changes in terms of strategy and in terms of implementation, (2) is open and so may be more palatable, (3) focuses on the future, not the past.

EXAMPLE—SUBSTANTIAL CHANGE

A computer software company was developing a new software product for the microcomputer market. The development approach was new for the firm. It was based on staff being assigned to the project from three different divisions in three different locations around the country. Without a tight deadline and technical challenges, this would have been enough for trouble. But, the project was cursed from the start with the organization divided, a tight deadline, and substantial technical hurdles. The issue was now why progress had slowed. It was clear that the deadline would not be met. People had proposed a variety of cures—ranging from new people to new software tools.

In examining this situation we can first see that our flexibility is very limited. Our scope is very narrow. We cannot bring in new people because that will slow down the project further. Latching onto a new tool will have the same effect. The scope is narrowed down to organization, project plan organization, and project structure.

Our strategy is to achieve parallel development effort so as to accelerate progress. The current project was hampered by central control. The regional managers were not committed to the new software product. The strategy had to address their commitment. The strategy was one in which the project was divided into three parts—regional managers were to manage their appropriate part. A central overall project plan was to control the separate plans. Integration and testing were to be done centrally with people from remote offices involved as well.

The changes were carried out in two stages. The first stage was decentralization. This got the commitment of the regional managers. The second stage was to establish the central control and management.

Did the project succeed? Eventually, yes. But it was late because the people in different offices did not get along with each other. If there had been more time, we could have suggested cutting down the number of offices and centralizing development to a greater degree. This is an example of how scope must include time.

EXAMPLE—MAJOR UPHEAVAL

In this example, a project has had several project managers. The project has several different focal points. No deliverable product has been produced in the six months of work. Seven people have been involved in the project. Morale is low. No one has a clear idea of what to do. Time for the revolution. It would take too long to go into the details of what was done. So we will focus on the highlights. The scope is wide open. The strategy is to assemble a team and get something done. So the first thing to consider is what reasonable milestones can be generated in a month or two that will be credible and not smoke and mirrors. The tendency is to throw out the entire project team. This is a mistake because there are typically some good people who possess knowledge and experience and are valuable to the project.

The first stage is to implement a new team and establish the short-term objectives. The second stage is to have additional milestones as well as an overall project plan. Also, we begin to identify tools and methods—for these we may almost totally discard what was previously used in the project.

This was also one of our more successful projects. At the start, no one wanted to work on the project. Toward the middle, everyone wanted to be on the project team. The project was also extended and expanded in terms of budget.

KILLING A PROJECT

But what if change does not work? Time to kill the project. Large projects are particularly difficult to stop. Witness the B-70 bomber, the U.S. main battle tank, application of early pesticides, and large public works projects. The project has received management approval—a mandate from heaven.

WHAT IS DEATH?

Death of a project is not only stopping work and reassigning people and resources, but also the process of getting as much out of the project in terms of what can be learned for the future so that we do not repeat the same mistakes. Death is accompanied by direct salvage. Death of a project involves many things:

- Understanding why the death of a project is necessary
- Seeing why it is so difficult to kill projects
- How to plan the organized death and mourning of the project
- How to carry out the death sentence most effectively
- What to learn from the process and the project

Many factors could have led us to this position:

1. Technology gap. A specific technology product was supposed to be ready and perform up to a practical standard.

2. Project management trap. The method of managing the project was cumbersome and ineffective. The project descended gradually into chaos and disorganization.

3. Disappearance of need. The original motivation and stimulus for the project receded, but the project was left high and dry on the beach. The project will yield a solution in search of a problem.

4. Project team discord. The project team works at counter-purposes. The project is in turmoil.

5. Loss of management interest and support. Management lost interest in the project.

6. Project growth and change. The original purpose of the project was lost in the shuffle. The project now is attempting to achieve new goals—impossible to achieve with the resources at hand.

7. Loss of support from the potential beneficiaries of the results of the project. These people don't want it now.

SITUATIONS WHERE PROLONGED LIFE IS NECESSARY

There are a selected few situations where keeping the project alive is necessary. Keeping it going may keep hope alive. You may want to keep the competition guessing. Think of the Strategic Defense Initiative (SDI). It successfully drained billions out of the defense budget of the Soviet Union.

ALTERNATIVE DEATH SENTENCES

Too little attention is given to this area. People often see it as black and white. It is not. The project, like the world, consists of many shades of gray and color. Here are some alternatives:

- A sudden death—The project is terminated through a single announcement and stroke of a pen.
- Gradual, slow death—Here the resources are gradually taken away from the project. The project dies a slow death. This is often politically acceptable because management does not lose face openly. However, this is often the worse method due to the impact on staff and other projects.
- Redirection and cutting losses
- Mothball the project—Freeze the project; lock the files and reassign the resources.

Sudden death at first glance has appeal. Management appears decisive. But it gradually dawns on people that no one thought through the impact.

Thus, everyone begins to scurry to pick up the various pieces. Gradual death through starvation presents a series of problems and challenges. Which resources are removed from the project? In what order? Removing a key resource may kill the project anyhow—suddenly. Removing bit players on the project team may have no impact. If you elect sudden death, you must provide direction to the project and a story to the project team. In redirection you may admit to the project team that there are problems and that the project as it is currently constituted is dead. The contingency plan is to both salvage from the project and to lose as little time and resources through reassignment. People on the project team are either reassigned or begin salvage work. Salvage is psychologically useful because people see something coming out of the project. Mothballing is only feasible if you really believe the need for the project will arise someday in the future. An example where this method is appropriate is when there is a technology gap that must be filled in order for the project to continue.

THE COST OF KILLING A PROJECT

Table 17.3 gives a list of various potential costs for you to review. How should you deal with the cost of death? The price of death should be put into the project plan and budget. This has several benefits. First, it makes everyone aware up front that the project may be killed. Second, by including tasks related to considering killing the project, we force the project team to address the reality of the project.

WHAT CAN BE SALVAGED?

We can salvage experience and lessons learned about the management of the project. This will tell us what not to do later. Some organizations

TABLE 17.3   Potential Costs in Killing a Project

Explanations and public relations in dealing with killing the project
Mothballing the project to potentially restart later
Keeping the project going to a logical stopping point
Paying off contracts and agreements
Reassigning and outplacement of project team members
Dealing with lower morale
Trying to learn from the project for other projects
Analyzing what can be salvaged from the project
Actual cost of salvage
Dealing with impact on other projects that will continue, but which depended on this project
Continuation costs for the project team members
Cost of storage and archival of project materials, equipment, and so forth
Tearing down what has been done on the project so far
Trying to come up with an alternative project to achieve the same project objective

even have employees train others how to avoid the same problems. We have the experience and expertise gained from using the tools and methods associated with the project. This can be invaluable in terms of providing guidelines and expertise for other projects.

THE SALVAGE PROJECT

What can we learn from salvage expeditions of the *Titanic* and Spanish galleons? First, they were well organized. They employed and still employ professionals who have done this before. This suggests that large organizations should identify people in different parts of the organization who are brought together to salvage a project.

Salvagers also have a very organized and methodical approach. For example, they first survey the project or wreck. Then they plan the salvage order. Underwater, if you screw this up, you could lose some critical items forever. Here is an exercise. From where you are sitting right now and reading this assume that there is a fire and that you have to evacuate—immediately. What do you take with you? What is irreplaceable? Experience shows that without a plan, people grab mundane items that are easily replaced. They leave behind irreplaceable records, photos, and mementos and regret it for the rest of their lives.

Salvage is then another project with these activities:

1. Survey the project to see what is available. This not only includes physical work and end products, but it also includes what is in people's minds. If the project team splits, the human record is lost.

2. Identify the information and material that may be lost first. This is often the experience of the project team. You might think that interviews can be conducted later. But the information will be tainted politically. This is why police conduct interviews immediately at a crime scene.

3. Identify people who will participate in the salvage effort. In diving it is not only the divers, but the support staff and apparatus that must be considered. Each person needs to be assigned a specific set of tasks.

4. Identify deliverable items from the project. After the initial survey of the project, a list of deliverable items needs to be presented to management. This accomplishes several goals. First, it helps in setting up the schedule for salvage. Second, it shows management the value of the salvage effort. Here are some guidelines from our past experience:

    a. Elapsed project time measured in weeks not months. The salvage operation must be rapid. Salvage projects reach the point of diminishing returns. When additional effort will yield marginal results, quit.

    b. Parallelism and rapid implementation. The salvage project needs to be carried out by the team in parallel. Otherwise, there is the chance that data may be lost and people may not be cooperative.

c. Salvage team meetings. The team should meet every day of the salvage effort. Progress is reviewed along with the results of the previous day. Plans are made for the next day.

HOW TO INTERVIEW PROJECT MEMBERS

Interviewing project members is very valuable to understand what happened and why. Not just for what really happened, but for what was perceived. Some suggestions on interviews:

1. Have the person state what he/she did during the project. What was their role at the start of their involvement? How did it change? You need to define a dispassionate sequence of events and activities to serve as the basis for later statements.

2. Get the person to comment on the tools and methods used in the project. This is very useful because it can be accomplished without emotion. It allows for a broader perspective. How did they use the tool or method? Did the method mesh with the tool? What tools and methods were missing? What could have been done to train and educate people more?

3. Now move to how milestones were reviewed and evaluated. This is starting to touch on the management of the project by focusing on deliverable, tangible items. Go into the process for feedback on milestones. How long did the review take? Did it consider content as well as presence?

4. Review the GANTT chart and template structure of the project. Did the structure make sense? Did the structure get in the way of the project? Was the project revised often? Were the revisions done in a logical manner?

5. How was the project managed? Who did you work with? Where did you go when you needed help? How were issues in the project handled? How were project decisions communicated?

6. What would they suggest as to how it could have been done differently? Have them vent any feelings now. Some people may focus on the process; others may focus on the work. They can start anywhere. You need them to comment on their thoughts about all parts of the project.

Who should be interviewed? Do not settle for the project manager and project team. Go after the people who worked on the project and left. Why did they leave? What signs did they see that the project was in trouble?

Contact the people on the periphery of the project. What did they know? Also, what did they not know? A source of difficulty in a project is often that the bit and cameo players were not provided enough information to do their job properly.

AFTER THE INTERVIEWS

The interviews will likely yield conflicting information. This is true with witnesses at the scene of a crime. How do you deal with conflict? Assume that everyone is right. Part of the conflict can be resolved by overlaying the factor of time. Some people may only have been involved in early stages of the project. The conflict remaining can be viewed in the context of perspective. Managers have a different view with different knowledge than people who were in the trenches.

WHAT CAN GO WRONG DURING THE PROJECT SHUT DOWN?

During the project shut down, many unforeseen events can occur. In one large aerospace company, employees on the team took home project documents and destroyed them. In another project a disgruntled employee attempted to murder the manager and salvage team. In yet another, the project team tried to stage a rebellion hoping to appeal to upper management through a coup d'etat. We live in a violent age—there is no reason to suppose that your project and organization are immune from the outside world.

Management should always have a mental list of potential events that may occur. Then you can look for early warning signs of problems. The salvage team can be listening and be aware. Here are some situations we have encountered and also guidelines.

The Disgruntled Employee

A person is exceptionally angry or upset. Or, an employee can be very quiet—too quiet. Have the salvage team focus on these people and get them out of the picture as fast as possible.

Lost Documents

Documents seem to have disappeared. Gather documents at the start of the salvage effort. Back up computer disk files before the announcement. Get the project files at the start of salvage.

Employees Who Want the Project to Continue

Employees complain that the death sentence should be commuted. Have an appeal or meeting process where employees can meet with a manager.

Dealing with Inevitable Depression

Conduct meetings to indicate what was done right and wrong in the project. Allow people a chance to vent their frustrations and feelings.

HOW THE ORGANIZATION LEARNS FROM FAILURES

How do you translate what was learned into action? This may be difficult because the findings may challenge standard beliefs of the organization. What did IBM learn from the success of personal computers? What did General Motors learn from the Saturn plant?

Where do you start? You first need to present the findings of the salvage effort and review. Also, prepare your implementation plan for change. Divide recommendations into tactical and strategic. Relate each recommendation to one or more findings. Prepare a table with rows of recommendations and columns of findings. Place an X in the table if the recommendation in the row is supported by the finding in the column.

For each recommendation you need to define the implementation steps as well as answers to the following:

- What will be the change, if implemented?
- What will be the impact, if things continue as they are?

Remember all changes have to be accompanied by a statement of what is expected as a result of the change. And you will have to identify how you will verify that the changes have been put into effect.

EXERCISES

For a specific project you know or have read about, define a list of changes. You need to see the impact of different types and levels of changes on the project. For resources assume that there were fewer people, more people, and the same number reassigned. For project scope assume that the project was cut back and narrowed. Seemingly small changes in resource levels can have a dramatic impact on the project itself.

You need to think through the process of killing a project. First, look around the house or apartment and think about where and how you spend your time. Where do you waste time? If you killed these activities that waste time, what would you put in their place? Now that you have this figured out, decide on the death sentence.

Move now to your work or what you have read in the newspaper recently. Try to find a project that has been killed because it ran into trouble. Why did it run into trouble? What would have happened if the project had been killed earlier? Who has or had a vested interest in keeping it going? Who benefits from its death?

EXAMPLE

The guidelines presented on the death of a project were employed in the project as part of the effort to terminate the old method of project management. A major effort that was discussed in an earlier chapter was the transition of staff associated with the old system.

The project underwent several major changes. This was in part due to experience learned in the earlier projects, the elapsed time of the projects, and changing technology. Over a period of eighteen months, two major system upgrades of hardware and software were undertaken. In general, change in technology was encouraged when there was obvious benefit and when there would be no negative impact on the staff's interface to the computer system.

Another source of major change was the manner of bringing new divisions into the overall project. Instead of waiting for a formal project kickoff, preliminary work began three months prior to the start of the formal project. Staff in the division was made familiar through demonstration of the system; they learned and saw it in operation with the first and the latest division to be using the system. This facilitated the transition to the new system in the department.

SUMMARY

Project change is almost inevitable in every project. Too many people treat it in a casual or panic-wrenching manner. What we have done here is to indicate the need and benefit of an overall strategy as well as an implementation strategy. The benefit of combining multiple changes at one time so as to reduce disruption to the project has also been emphasized.

Killing a project, determining the method of death, and carrying out the sentence are important not only for other projects, but also for the project management process. Through failure and death we learn a great deal about how to attain success in the future. You should have gained an understanding of the various steps in killing a project—topics that people give little thought to on a daily basis.

18 Lessons Learned from the Project

The review of a project is often called a postimplementation review or postmortem. Because we associated death of a project with failure, we will adopt postimplementation review here for a general project. The bottom line in this chapter is to discover how we can gather and use lessons learned from a completed project to help us with current and future projects.

WHY LEARN FROM PROJECTS?

The review is a way to build cumulative knowledge for later projects. It is also a way to give the project team feedback and provide an opportunity to share experiences. Experience and knowledge of tools and methods can be nailed down. The review can also identify opportunities for improvements later.

WHAT CAN WE LEARN FROM THE PROJECT?

Some people view history in terms of results. The ends justify the means. Others are caught up in the process and neglect the outcome. We want to avoid both extremes. We need to review the project from the standpoint of process, organization, technology and methods, resources, and users of the completed project.

The dimensions of what we can learn include the following:

- Project management process. This includes how we manage, control, and plan projects.
- Project management tools and methods. These are tools and techniques that support the project management process.
- Tools in the project. These are the tools used by the project team in doing the work.
- Methods in the project. These are the methods in the project as well as the relationship of the tools to the methods.
- Organization. How did the functional organizational units support the project?
- Training. How effective was the preparation of the project manager and team for the project?
- Project structure. The organization of the tasks and assignment of resources to the project.
- Quality of project results. This can be viewed from several perspectives: the end user of the results of the project, the method, the tools, the project manager, administration, and so on.
- Budget perspective. Was the project on time and within budget? What was the rate of expenditure?
- Specific line organizations. The perspective of the project from that of specific line organizations who were involved in the project.
- Methods used to resolve conflicts.

Why have this list? Because you want to be as inclusive and organized in reviewing a project as possible. Some amplifying comments are as follows:

- The process of project management has to be considered with the project leader removed from the analysis. Otherwise, this can turn out to be a review of the person and not the role.
- There is a sharp distinction between tools and methods as there is in the rest of the book. The tool (regardless of whether it applies to project management or to the work) is placed within the context of overall methods. Tools fail when they are not connected to the methods. Methods, on the other hand, cover a wider area.
- With respect to the budget, it is not just looking at the total resources, but also at the expenditure rate and distribution of costs and resources.

- A great deal has been written about the functional versus the project organization that was mentioned in an early chapter. There are many differences and points of potential conflict between these. Examples abound of conflict in criminal task forces (project), military special forces that cross the service branches, product management versus standard line organizations. At the end of the project it is useful to determine how this conflict manifested itself and what could be done to address the problem.
- Quality. Obviously, we can finish a project, but the work can be substandard. The world has seen many buildings and bridges that collapse because of faulty construction during the project.
- With respect to project structure, we want to ensure that the structure and its management were logical and that changes were made consistently.

WHY DO WE OFTEN NOT LEARN FROM PROJECTS?

Project reviews are sometimes witch-hunts. That is, the organization perceives a need to find a scapegoat for problems in the project, especially if the outcome was not good. In democratic elections, this makes sense because the people, who are thrown out, were the ones in charge. Project witch-hunts are not as nice. Often, lower level staff and team members are blamed so that management can be absolved of guilt and association. More generally, people do not want to learn from the project, but instead want to achieve a political goal through the project using history as the means.

Another reason we don't learn is that we pretend we are too busy on other projects to learn. Many of us do not wish to face unpleasant facts. This is one reason why methods that have proven unsuccessful are still used. The classic example in World War I occurred in the trenches on the Western Front. It took over two and a half years before the army admitted that the traditional method of warfare was obsolete against machine guns, barbed wire, mustard gas, and trenches. When tanks were massed on a large scale, the German front collapsed. Yet the tank was ready to go in 1915. Why does this happen? People often have boxed themselves into a corner (specific position) and no one offers a way out. A related reason is that people often tend to accept the process as sacred so that they do not question or try to improve the process. People keep doing things the same way even when they fail. Many churches in medieval Europe literally collapsed because of the same faulty engineering. The problems were known, but were not acted upon and so were repeated.

HOW TO SPOT A WITCH-HUNT

You must be able to spot a witch-hunt so that you do not get caught up in the hunt. How does it start? A manager may show up at the project meeting and announce that an audit or review of the project is to be conducted by an outsider. This can be a consulting or accounting firm. Listen to the words carefully as the charter and mission of the audit are described. Are they specific? If they are, then there is hope that this is a sincere review. But if it is a general review, watch out! The game is afoot!

WHAT DO YOU DO IF IT IS A WITCH-HUNT?

Let us assume that you are on the project team and that you suspect that the review is a witch-hunt. What do you do? After the initial meeting ask the manager what the focus of the review is and what materials you should prepare. If you are met with evasive answers, this reinforces your feelings that there is a hunt. After the meeting wait for the people to come to you. Organize your file and materials and be ready. Review the project in your mind in terms of major events and problems. If you are having trouble finding problems, then concentrate on the politics of the project. Ask yourself the following questions:

- Did the project antagonize managers because it was completed?
- Who stands to lose politically if the project is viewed as a success or failure?
- Did the project manager make some people angry?
- Did the project take resources from competing projects? If so, how were they impacted?

When the people who are conducting the review visit you, be cooperative. Show them what you have and explain your role carefully and precisely. Do not act shy or reticent. It will convey that you are trying to hide something. If they want to take documents from you, indicate that you still are using them, but that you will make copies.

What kind of questions are you being asked? Are you being asked how certain people behaved during the project? Beware if they are trying to take you into "their confidence." This is very similar to police interrogation with multiple suspects. You are a suspect. They want you to tell them about the project. Be careful.

As the hunt continues, other project team members will be getting nervous. They may ask you what is going on and what you have heard. Play dumb and do not spend a lot of time with them unless the project requires it.

Let's look at the review from the standpoint of the reviewer. The reviewer was brought in to find something. If nothing is found, the reviewer may be in trouble. The manager is at fault for bringing in the reviewer. After all, the reviewer may have already been told what is to be found. The only thing that remains is to find evidence.

WHAT GETS IN THE WAY OF LEARNING?

Enough of this negative subject. Just keep in mind that it may happen to you. Be prepared. There are factors that keep us from learning from the project and implementing change. Here is a list.

- What should be done with the findings of the review? In many situations there is an awareness that the process cannot be changed easily. Management may have endorsed and be committed to the process. This can be overcome by stressing that any process can be improved.
- There are vested interests in the tools and methods. Reputations of people may be riding on the effectiveness of a specific tool or method. This is true of many computer software design methods. What happens to the method? Eventually, the method is discredited and not used. Lip service is paid to the method, while people go about their work. This happened in China during and after the Cultural Revolution and in reaction to the cult of Mao Tse-tung.
- People who are assigned to new work and other projects have to accept the tools and methods of the new projects. If these are the same as those of the old project, there is a reluctance to criticize them. What should be done? Have the review started before the project is completed and have it finished while the information is still fresh.
- On large projects that have spanned several years, there are memory problems. Who can recall all of the details? Addressing this situation falls on the shoulders of the reviewer.

WHAT IS PROJECT SUCCESS?

Project success is a vague term. A project could be a physical failure, but a political success (e.g., Aswan Dam). This indicates that whether a project was a success depends on the perspective from which the project is viewed. However, we can give some criteria for success.

PROJECT ON SCHEDULE AND WITHIN BUDGET

Was the project completed within schedule and budget? Answering this question is not as simple as it seems because the budget and schedule may have been changed many times. The answer requires a look back across the project.

END PRODUCT IN USE

Is the end product of the project being used? This question goes to the heart of the project. Later, we will discuss issues of quality, maintainability, and completeness. Here we ask if the results of the project are being used in everyday business processes. If they are, then there is some measure of success.

PROJECT MANAGER AND TEAM PERFORMANCE

How well did the project manager and project team perform? This is not a simple rating as in the Olympic Games where judges hold up signs with scores. Did the project team deal with issues early or as soon as they surfaced? Or, did they fester and get worse? Was management kept informed about the project? What signs were there of misunderstandings?

In general, most projects are viewed as a mixture of failure and success. Some things worked; some did not. The end product is often not quite right. It works but unforeseen behavior and impacts occurred. A prime example of this is a new freeway or rapid transit system. At the start of operation the transportation system is a success. Transit time is reduced. But now people start to move to be close to the system to shorten their commuting time. Transit times increase due to clogged surface streets. This happened with the BART (Bay Area Rapid Transit System) in San Francisco, California. The system works. But maintenance is a constant battle. More cars cannot be ordered because the company that made the cars went out of that business.

Thus, a key ingredient of success or failure is not the short-term success or failure, but what happens in the longer term. This relates to operability and maintenance. There are two automobiles that sold quite well. Sounds like success, eh? They were not. In one case, to change the front shock absorbers, the entire front end of the car has to be taken apart. Very expensive. In the other case, to do a tune up, the engine has to be lifted from the engine compartment. Why? The engine has to be removed to reach the rear two spark plugs because there is not enough space to work in under the hood. Sales of both later declined. We learn from this discussion that there can be both success and failure. Furthermore, the situation can evolve over time.

LOOKING AT THE PROJECT FROM DIFFERENT PERSPECTIVES

To perform a successful review, you have to consider how the project looks from different points of view.

- End user of the final product. This is the driver of the car on the freeway—the ultimate user of the system.
- Operations management and staff. These are the people who have to make the system work after it has been completed in the project.
- Project manager and team. This is the view of the processes of the project from the inside.
- Management. This is the external view of the project processes.
- Tool and methods. How did the project take advantage of the tools and methods? Did they learn from the tools and methods as they were employed?
- Other projects. This is also an external view based on the dependencies of other projects to the completed project.

Other perspectives are possible. However, you should at least consider these.

HOW TO CONDUCT THE REVIEW OF THE PROJECT

Now that we have paved the way for the review and its importance, where do we begin? We start in a way similar to the salvage project of the previous chapter. We need to gather information on both the project and the resulting system. Some additional comments are appropriate for the internal project process.

1. Build a timeline of the project in terms of milestones and key events. Show key decisions on this timeline. This timeline can then be verified by the project team. It can be used as a reference. Key events include people coming and going from the project; milestones; decision points; crises; major meetings; approval or rejection of budgets and milestones.

2. You need to assure the people that this is not a witch-hunt. How? By defining very clear objectives and issues that you seek to address in the review. But that is not all. You also have to define how the results will be used.

3. You will have to gain the confidence of the team. Gather information and do not share information in confidence with the team. Tell them that anything they say will be public knowledge. You are trying to build consensus as to what happened, why it happened, and what might be done in the future. You need to have an organized approach for your work. We have already pointed out the need for the project file and for interviewing in a structured way. Here are some other suggestions:

4. Build a list of tentative findings on lessons learned as statements. These have to be complete, organized thoughts. Label these in terms of major and minor and also in terms of what they relate to in the project.

5. Build a set of tentative recommendations. Each recommendation should follow from one or more findings. A recommendation can apply to current and/or future projects. If you cannot relate a recommendation, then drop it. You are trying to impose your own feelings on the report. The recommendation should be followed by a statement of benefits.

6. State findings and recommendations in very dry words that are not political. Otherwise, they will color the report unfairly.

7. Start building a third list—the list of actions to be taken. Each action should relate to one or more recommendations and should be followed by a discussion of the benefits of the specific action.

Remember that you have to assume that motivating change is a hard sell. Knowing the battle is uphill, you have to build benefits that are as tangible as possible. Also, you have to define realistic actions. What is a realistic action—something that does not require money or disrupt the organization.

Here is another idea. Review the cumulative, building lists with the project team. This will gain their involvement and, hopefully, their commitment. The project team is important in that they are the ones who will be asked if these changes would have made a difference in the project.

REVIEW OF PROJECT'S END PRODUCT SYSTEM

We will refer to the "system" as the end product. Because we are considering many different projects, the system can take many forms. Let's first consider what we could gather from the system. A detailed list of questions about a system is given in Table 18.1. Answering these questions depends

TABLE 18.1   List of Questions about Project Results

Did the system meet the original objectives?
Did the system meet the latest approved specifications?
Did the system adhere to the design?
Is the system being used?
How is the system being used?
What is the effort required to operate the system?
How is the system being managed?
How is the system being maintained?
How does the system interface to other systems?
What are the cost and effort to operate the system?
What are the cost and effort to maintain the system?
What are the cost and effort to enhance the system?
Is the system flexible enough to support change?
What are the changes to interfaces with other systems, or do these match those designed
   and built in the project?

on how-long after completion of the project the review occurs. Even so, you should be able to gather information any time after completion. We can also ask more broad questions. How does the system compare with those of other organizations? This assumes that the system is not unique. Of course, everyone thinks of their system as unique. But you can, with little effort, find parallels with other projects.

How should you gather information? Direct observation is one way. You should observe the system at different times and under different circumstances. You can interview the people who manage it, operate it, and maintain it. Under certain circumstances, you can talk to the users of the system. You have to be careful. In some cases, if the review is conducted poorly, people will modify their behavior and tell you what you want to hear.

This means that you have to be as open in collecting data on the system as on the project. How can you judge yourself to be successful? If the information yields suggestions for further improvements. If people tell you that it is fine as it is, then you may begin to assume they are not using the system.

How should you organize the information from the system observation and interviews? You should prepare a list of findings. Each finding should be accompanied by the source and evidence supporting the finding as well as any impact of the finding. You should also develop a list of recommendations for future enhancement and change. Each recommendation needs to be supported by benefits and action items for implementation. Also, you need to develop a list of issues that need to be addressed. Each issue should include a discussion of what will happen if the issue is not addressed.

NEED FOR A STRATEGY AND STRATEGIC VIEW

The tendency in these reviews is to summarize the lists we have identified. Go to management and present them. Everyone will appear to be satisfied because you came up with results that the project team and people involved in the current system approve. It is not enough. Remember you are trying to carry out change. You need more than lists and tangible benefits. You need an implementation strategy and strategic view of the project.

What do we mean? First, a strategic view of the project management process will relate all of the changes you are proposing into a whole. How do you express the whole of the changes? Use a scenario. Tell management how things would have been different and how they could be improved overall. This will not only help them understand, but it will show that the recommendations are interrelated. Otherwise, if they implement actions 1, 3, 6, and 7, they may fail because they did not do 2 and 4 (forget 5).

A second part relates to similar projects in the future. You must have gained information on the current system that will help in similar projects in the future. Again, the scenario approach will work. Note that the scenario approach can be compared to the television advertisement for headache remedies. During the ad, the person is shown with no pain. This is the scenario.

STRUCTURE OF THE REPORT

Table 18.2 presents several alternative outlines for postimplementation review reports. One focuses on action for future projects. Another attempts to be more of a factual summary with no immediate call for action. You should be able to use both of these. Sometimes after a review is conducted it is decided that no major changes are required—only some minor work and revision.

The report should be less than 10 pages and be list oriented. An appendix of the report should produce mappings of findings, recommendations, and action items. People cannot be assumed to automatically relate these. You also have to develop the implementation strategy in the report. Without this, the recommendations and actions will hang and be picked apart.

In your report you need to clearly state the process you followed in collecting the information and doing the analysis. Lists of people interviewed and contacted should be given. This will show organization and

TABLE 18.2   Alternative Outlines for Postimplementation Review

A. Summary approach[a]
   1. Purpose and scope of review—approach in doing the review
   2. Review of the system—review of the system that resulted from the project
   3. Project summary—highlights of what happened in the project
   4. Findings—specific findings related to the review
   5. Conclusions and recommendations

B. Action approach[b]
   1. Purpose and scope of review—similar to the first approach, but his emphasis on actions and issues
   2. Specific questions and issues—identifies specific questions and issues to be addressed in the review
   3. General system review—similar to above, but with an emphasis on the issues of section 2
   4. General project review—similar to above, but with an exmphasis on the issues of section 2
   5. Issue analysis—analysis of the individual issues
   6. Conclusions
   7. Recommendations—specific recommendations for action with reference to the issue analysis of section 5

[a] This approach focuses on the system first and then on the project. The review document is structured in a top-down manner with specific findings following a general review. The approach is passive.

[b] Note that this review is conducted with a specific agenda of issues in mind.

support the credibility of your findings. The recommendations should follow naturally from the findings. Otherwise, the gap creates doubt about the recommendations. A similar argument holds for the actions.

You may wish to identify alternative actions. You could list actions requiring no money or organization change. These could be implemented more quickly. Another group would consist of actions that require more effort.

HOW SHOULD YOU SELL YOUR REPORT?

You will need to convince management of your recommendations. This goes beyond the report. Move to the presentation. Let's consider a traditional presentation. You would stand up and state the findings, followed by recommendations and actions. Very dry and very dull.

A more interesting approach is to divide the presentation into three parts. The first part refers to the system itself produced by the project. This can be in the format described above. Now for the review of the project. Present the findings along with someone from the project team to reinforce what you have said and what problems were created. This should warm up the audience for the recommendations and actions. Again, avoid being dull. Move to your strategy and scenario. Here is the hope for the future. Give examples of benefits. This approach has proved to stimulate more management support.

EXERCISES

Review the newspapers and magazines for articles that cover completed projects. Then go to the library and search for additional articles on the same projects and systems. This will show you how interpretation can change.

Now move closer to home. Look at a recent project you completed around your home or apartment. Conduct a review of your work. This can be very unpleasant. But, you will learn from it. For example, take a household plumbing project. Did you start the project and then have to stop because you lacked a critical part? Then you went to several stores to find the part—eating up time. Did the elapsed time for the project take too long? What could have been done in terms of organization and planning at the front end of the project?

You should also develop skills for building scenarios. A scenario is a model of how the world of the project would change if specific recommendations were followed and implemented. A scenario integrates the recommendations and applies them to a project. You have to be able to do this well. Otherwise, you will have less success in getting your ideas across.

The general idea for a scenario is to take a set of recommended changes and apply them to the work flow of a project. How would people have done things differently if they had followed the recommendations? To develop the scenario you first have to define the work flow in terms of specific activities. After doing this, you can now develop a set of recommendations. Next, attempt to proceed through the work flow following the recommendations. Are there gaps where no recommendation applies? Then you may wish to expand your recommendations. Perhaps, you missed a finding. Continue refining the strategy, scenario, recommendation, and work flow until they are stable.

EXAMPLE

Because our overall project was composed of a number of middle-sized projects, it was possible to learn from one project and apply the results both forward to later projects and backward to earlier projects. One thing that was instilled early in the project and continued throughout is that reengineering and process improvement are continuous and not one-time events. That paved the way for applying lessons learned from other projects.

In addition to improved methods, rules, and use of tools, a fundamental lesson learned was how to organize the later projects to be shorter in duration and easier to implement in terms of organization change. After the first three divisions, the approach was to identify the likely organization and personnel issues at the start of the project based on the experience of the earlier projects. This not only improved the schedule, but also reduced the uncertainty among the staff in the division.

Another lesson learned was that the results of the review on the process and the results of reengineering and the new project management approach needed to be disseminated widely and almost demonstrated. This not only made people aware of the results, but also worked to widen participation among managers.

SUMMARY

The importance of reviewing a project cannot be underestimated. Otherwise, how will the organization learn and improve? Moreover, the review is very cheap in effort and hassle. There are few resources needed except for access to people, files, and systems. As we said before, the reasons for reluctance to do this often stem from political factors. The example that we have been using points to a continuous measurement process as opposed to a specific report. This can be implemented through the review indicated in the chapter, together with the issues approach in an earlier chapter.

19 Where to Go from Here

PROJECT MANAGEMENT REDEFINED

As you have seen, project management is a thinking process. It is not as simple as GANTT, PERT, or CPM. Project management applies organization and structure to accomplish the work. Dealing with issues, conflicts, and opportunities is a major project management activity. It is not a question of whether to use project management, but instead which methods, techniques, and tools are most appropriate to the project. Selecting these correctly and maintaining flexibility are key concepts covered in this book.

THE VALUE OF PROJECT MANAGEMENT

The current decade has been seen as more turbulent, with faster paced change and uncertainty when compared to the relatively calmer periods of several earlier decades. If there is more change, why use project management, which focuses on structure? Change and the requirement to achieve results quickly with fewer resources are supported by project management. Project management is a comprehensive tool for us to

- Apply standard methods and structure to a situation;
- Deal with changes and impacts to the project;
- Learn from past projects for the future.

Project management supports cumulative learning and experience. Otherwise, we might view a series of different projects and work as totally independent, when, in fact, they share much in common in terms of organization issues, project team interaction, and project change.

USE AND ABUSE

Proper use of project management not only gets the work done better, faster, or with more control, but also provides a common perspective across many different projects. Without a basis of comparison, we are left with budgets and money—not very useful or appropriate in many situations.

But project management can make a situation worse. In each chapter we have pointed to pitfalls and potential factors for disaster. Let's examine some of the more common ones again. Consider the project where the project management tools and controls are applied unevenly. A disaster is in the making. If there are no controls, the work can get out of hand. If there is an emphasis on tools, there is far too much effort spent on the tool—learning the tool, telling everyone about the tool, using the tool, becoming proficient with the tool, and educating others on the tool. The process consumes the available time and there is no end product. Or, how about the case of the project manager with a small project mind set who is managing a larger project? The manager drives himself or herself and members of the team crazy by attempting to overcontrol and impact every task. Problems are not viewed in proportion. The forest cannot be seen through the trees.

HOW TO PREVENT PROJECT MANAGEMENT ABUSE

First, be flexible. Don't decide on the tools, methods, controls, and techniques right away and then be rigid. You have to adjust the elements of project management to meet the needs and facets of the situation. When things are going well, you might employ the tools and techniques one way. Our general example employed this approach. That is why the manner in which tools were used in a specific situation were not documented in stone; instead, there were guidelines in the use of the tools and methods. When there is a crisis, you focus on the problem areas. The only things that tend to be fixed are the project team members and the reporting structure as well as the project.

Second, view flexibility, uncertainty, and change positively. Suppose that you are a project manager and someone proposes a better way to do something. You know that this suggestion will impact the plan. So you respond that this is a good idea, but it would take too much time to change the project. You are getting into trouble because you are dismissing initiative shown by a team member and reinforcing the status quo. Being a successful project manager means dealing with constant change and unknown factors.

SCALING PROJECT MANAGEMENT

Flexibility is related to how we adjust project management to the project. At the early stages of a project, we might organize project management informally. Later, as the project grows in scope and complexity, you expand the controls and methods. At the end of the project it can be scaled back again. The bottom line is that project management should reflect the project and what is happening in the project at the time. In our example, we began with a very large level 4 project. This then became the vision and was replaced in detail with level 2 and 3 projects. Several times a project moved between levels.

PROJECT MANAGEMENT AND THE ORGANIZATION

Keep in mind that there is natural conflict in many organizations regarding project management. Our overall example pointed out several such conflicts. Projects and project teams often cross departments. There is a constant source of strife between the project and the line organizations. Some line managers still believe that they could attain the project objective without the project. They view the project as another level of bureaucracy. In turn, project managers frequently view line managers as barriers and obstacles to getting the work done.

This natural tension is positive in that each side looks after its own interest. The point here is that the project manager and the project team have a stake in the project's success. To a line manager, the project might seem like just another project. The project approach provides for focus and commitment by the manager and team.

Project managers need to recognize that conflict exists and is a natural part of the process. Therefore, the project manager must anticipate problems and potential conflict. The project manager also should make an effort to establish good relations with key line managers at the start of the project and continue to maintain communications throughout the project.

WHY BEING A PROJECT MANAGER IS A GOOD IDEA

To some, the duties and responsibilities and especially the risks appear overwhelming. But there are rewards. First, being a project manager offers a person not currently in management to demonstrate their abilities. Waiting to be promoted may take much longer. A project manager has a better chance for visibility with upper management.

A second reason is that the project manager gets the opportunity to work with different departments and their needs. This is a "free, no cost" way to get exposure to the operations of various departments. Combining these two points, project management can offer the later opportunity to move into a line manager's job after the project is completed.

Promotion in the future will rest more on achievement and a track record as well as the normal contacts and networking.

Success in a project and, especially turning a troubled project around, provide these achievements. By the way, the project manager in our example project was promoted two levels at the end of the project.

Consider in contrast, two resumes. One is for a project manager who successfully managed three projects over a 2–3-year period. The person can point to specific accomplishments that are tangible and measurable. The second résumé is for a middle line manager who directed 40 people in day-to-day work. Aside from management and achieving some normal business goals, there is nothing exceptional in this second résumé. Which person would you hire?

GETTING ON THE RIGHT PROJECT

The right project for you is not one that is already rolling along and will be successful. There is less chance to make your mark. If you have a choice, look for new projects and for projects with some risk. The project, however, needs to have some chance of success.

What are some other characteristics of good projects? They should not be too large. You may get buried in process. The project should not extend much over one year. Otherwise, you may get worn out and the project might drift, as we have seen from examples. The Taj Mahal project lasted 22 years, but it had continuous management support from the same person. Not as likely today.

What if there are no projects available? Look around for situations where a small project could accomplish some specific benefits. Our favorite and one that we recommend is a project to reengineer a business process and its procedures in a department to increase efficiency and effectiveness. Many organizations have ignored their underlying processes that keep the business afloat and concentrate on new ventures and jazzy opportunities. You should focus on the mundane and bread-and-butter processes.

What about a project that is in trouble? Study the project and determine why it is trouble. If the problems are outside of the project, it is less likely that you can succeed. If, however, the problems are internal to the project and the project team, then this might be an opportunity. How do you volunteer? After learning about the project, formulate some ideas on how the project might be improved. Then go to a manager whom you trust. What is the worst thing that can happen? You will be told no. Not bad, but at least you got some recognition as someone who volunteers and who thinks about the company. What else bad can happen? Your ideas may be stolen. Don't worry if this occurs. You still will be known by the manager(s) as the person who came up with the ideas. However, if you do this several times and your ideas are always used without credit, it is time to polish up the résumé and start looking for other opportunities. You know what? Even then it works in your favor because you can identify these ideas when you interview for another job. That the ideas were put into play is evidence of the ideas' viability.

GETTING STARTED WITH MINIMUM COST

We have given you a number of suggestions as to where to start using project management methods at home and at work. You need to be very conservative here. If you attempt to do too much and push too far too quickly, you will more likely become discouraged and then give up. Project management takes time to learn and do. The payoff from using project management does not come instantly, but instead later in the project as problems, opportunities, and issues are encountered. We don't want you to get depressed and quit. Concentrate on the human skills of being a project manager or a project team member. Avoid tools and formal methods at first. These take more time to learn.

LEARNING FROM THE PAST

We do not want to repeat what we stated earlier. From all of the historical examples we have used, you can see how we can learn from the ancient past. It is now your turn to learn from the recent past and the news. To test yourself, select one subject that is written up in several magazines and newspapers. Read the articles and compare their meaning and tone. What are they trying to say? How is the event being interpreted? In theory, the more time that passes, the more interpretation there is, until a consensus is achieved. Some events such as the Kennedy assassination live on, in part, because of the lack of consensus.

What can you use from the past? First, pursue the thinking process of how a situation was addressed. Second, consider the actual data and details. This is often useful in helping you remember the event a year later. A picture, an image, a fact, or some specific detail often provide a memory trigger for remembering.

OBSERVE, OBSERVE

The easiest way to begin thinking about project management is to observe it around you. Think of events in your news as being part of projects. Start observing, evaluating, and learning from events as being project related. Keep a diary of what you observe. Keeping notes provides a structured way of observation. Always keep your notes to yourself and never reveal that you keep a diary. Date each entry in your diary. After you write a sentence, never go back and change it. You want to write it down as you first remembered it. If you change your mind, then write down the new interpretation later and refer back. If it helps for reference, use a different color ink for different types of meetings or thoughts. A friend of ours uses black for normal meetings, red for problem meetings, and green for budget meetings.

Observing within the Project

You are a member of the project team. Be observant. Watch how the project control and management process work. Does the manager keep in touch with the project team? Is there good communications with line managers and organization? What could be improved? Again, keep a diary of your thoughts. When you have an idea that something is going wrong but this is not recognized, write it down in your diary. You are testing your ability to detect symptoms of problems. This can help you and the project as well.

If people in the project take you into their confidence, there is a tendency to press for action based on this information. Don't do it. You betray the confidence. If you are forced to suggest action based on this information, give the credit to the person who provided you with the information.

The Importance of Self-Evaluation

Observation is an important part of overall self-evaluation. In self-evaluation for project management, you want to improve your skills for the future. It relates back to our chapter on communications. First, evaluate what you heard and observed. Was it complete? Did you observe key events and any turning points?

Second, assess how you analyzed and interpreted the information. What did you overlook? What did you miss? What did you interpret correctly and incorrectly? In answering these questions, review the order and thought process you employed.

Third, analyze what you did with the information and the analysis. Did anything improve? What did you do differently? Go back and look at your diary. Take two meetings or dates where similar types of items were discussed. Compare the results of these meetings.

HOW DO PEOPLE VIEW YOU?

Let's assume that you are a member of a project team. How are you viewed by the team? We do not mean their opinion of you personally; we mean what role and capabilities do they perceive you to possess. Are you viewed as a technician? As an expert on a certain area? As a person who does a lot of work? As a problem solver? Do they ask for your input? You probably want to have them view you differently. Don't kid yourself and want to be perceived as a manager. Concentrate instead on being seen as a problem solver. A problem solver is a person who can quickly get to the heart of a problem. This person can then build consensus about the statement of the problem. Next, the person can formulate alternatives and determine their fit with the political realities of the situation. It will be easier to concentrate on the first part—defining the problem. Can you focus on a problem and attempt to understand its ramifications?

BEING A TEAM PLAYER

While we have spent time on project managers, most of the time we are a member of a project team. You should keep in mind your obligations to the project team. Being a team player is often a hard and thankless task. It is not just doing the tasks you are assigned to the best of your ability. It is also participation in the identification of issues and opportunities. It is the initiative to suggest possible approaches and solutions. The reward for this effort is not just in helping the project, but also in helping build the team and in increasing communications.

There are situations where you will be torn as to what to do. Suppose someone shares something in confidence that will have a negative impact on the project. It is too important to sit on. You need to do something. First, attempt to verify the information. Look for other symptoms and signs. Second, think through what has been told to you in terms of its impact on the project overall. If the likelihood of any impact in the near term is small, then wait. If it does have an impact, then you might go to the project manager and indicate what you have learned. Keep the source confidential, otherwise, you will have betrayed a trust that may not be extended again.

FOCUS ON THE METHODS AND THEN THE TOOLS

People are attracted to tools. They are offered by firms through marketing. Only the best features are shown. In contrast methods appear blah and drab. Methods force you to think about the process and the steps. These are some of the reasons why tools drive methods for many people. The reality is that the stability and attaining the result are due to employing the method. The tool works in support of the method. That is why we say that the emphasis should be on methods and not tools. Methods should be more stable than tools. Why? Methods are tied to the business process; tools are tied to technology and the press of marketing by the maker of the tool.

When getting going in project management, you should first identify and establish your methods. These may initially be taught through education and reading, but they are honed, reinforced, and enhanced through experience. The methods you apply through practice will be with you throughout your life. Take an example. Suppose you want to add a series of six four-digit numbers. How you add these is based on what you learned in school. If, on the other hand, you use a calculator, then you are forced to change your method and learn how the calculator operates to enter the numbers. The method surrounding the calculator depends on the specific make and model of calculator.

STANDARDS

The above example also shows us the importance of having standards. We have employed standards, conventions, and rules in the book in many chapters. Standards are important because of the need for uniform reporting and the capability to aggregate project data. But there is another reason. Much of project management detail is tedious. In doing tedious things, sometimes our minds change the method for variety. People who write by hand often may switch between upper and lower case in printing, for example. Standardization can help minimize the tedium and effort since we can set up canned work breakdown structure, resources, codes, and so forth, that can reduce the time to set up the schedule.

How do you acquire standards? Look around you and identify conventions and standards used by others and in other projects. Start building a list of these. Try to understand why they are used. How do they promote work on the project? Do they facilitate the work? Over time you will start to develop your own ideas on standards. Some people tend to suppress this because the place where they work stresses compliance. You should not be dissuaded. If you have your own standards, then you will tend to follow these instinctively.

SOME THOUGHTS ON TOOLS

We have been pretty hard on some tools and their abuse. On the other hand, we have pointed out a number of areas where tools are very useful. Project control and the issues database are two examples. There is nothing like a good tool. It can reinforce the method. If you get proficient with it, you will do quite well. Beware! Do not get caught in the trap of being regarded as a toolsmith. Then you will be regarded as the techie nerd who knows the specific tool. Instead, keep some of the expertise to yourself. If someone inquires, then respond that you are familiar with the tool. Also, you can volunteer to help others in the use of the tool. People tend to set themselves up as experts because they exhibit their skills. Word of mouth then attracts attention. They, in turn, enjoy the attention and devote a great deal of time to the tool.

The toolsmith has another problem. What if a new tool comes along? Is the toolsmith obsolete? Will there be a new toolsmith for the new tool? How can the toolsmith gain proficiency in the new tool while maintaining the existing tool? With technology change, being a toolsmith can be hazardous to your longevity and health.

Toolsmiths also tend to propagandize their tools. Some not only offer their tool, but use the tool to attack any problem that comes along without thought as to appropriate fit. Sounds stupid, doesn't it? It happens all of the time. A child who learns how to use a hammer wants to hammer everything. A teenager who learns to drive, wants to drive even the shortest distances.

Getting the Right Level of Proficiency

We have already said that you should not attempt to become an expert in the use of specific tools. This may typecast you as an expert—short-term security and recognition and long-term career risk. Tools change for several reasons. First, there can be new versions of the tool. Second, the tool may be replaced by some better one. Third, your organization may discard the tool and adopt another. So you do not want to put too much time into learning a tool.

You need to know enough about a tool to use it effectively in a project. How much is needed? If you are learning things that will assist in analysis or presentation, then this is probably useful. If you are learning nuances and shortcuts, then you may be spending too much time on the tool.

————— DO IT YOURSELF

In the past, we would have to suggest strongly that you do more of your work yourself. Today that is less necessary. The wonders of computers and communications mean that, like Lucille Ball ("I Love Lucy" television show) and the vacuum, we can do the work ourselves. In the series she was sold a vacuum cleaner. To get the money for the vacuum, she reduced her cleaning service budget. She ended up doing the work herself with the vacuum.

Because you are going to do more for yourself, you will have to learn more tools. This is another point supporting our position that you cannot afford to learn a tool with great proficiency because you have to learn many tools. In an average office, you might have to learn

- Electronic mail and Internet
- World Wide Web
- Electronic facsimile
- Electronic forms
- Word processing
- Spreadsheets
- Group scheduling and groupware
- Fax machine
- Copier machine
- Voice mail
- Typewriter

This is much more technology than was the case 20 years ago. Moreover, in the past, you often took the document to someone who used the technology. Now you are on your own. If you look like a fool in using it, it looks bad for you. So, you need to gain some knowledge and expertise—but not proficiency.

Because you are going to do more for yourself, you need to be better organized. Set aside time for testing your knowledge and abilities with the tools. You have to use the tools so that you won't get rusty. Try to integrate these tools around your work habits or methods as we suggested with project management software.

YOUR WRITING

Do not write too much. Some people we know write up problems, status reports, and almost anything. The communications channel is getting cluttered. People see so much from you that there is little impact—even if you are pointing to a disaster. The same applies to electronic mail and voice mail. These systems are often easy to use and enticing. They lure people in to use them. The technology people want you to use them because it helps to justify their positions and jobs. We suggest that you read everything you get, but that you keep messages and letters to a minimum. They will have a greater impact.

On electronic mail, keep in mind our suggestions on writing electronic mail messages. If you are in an organization where you can subscribe to a variety of different electronic mail lists, then you should, by all means, subscribe. It doesn't cost anything and you will learn a great deal about what is going on in the organization. When you run across good information, print it out and save it.

Be your own worst critic. That is, you need to be able to critically review what you are writing and doing. In the old days, you could write it by hand. You could depend on a secretary to clean it up somewhat and give it back to you. The elapsed time gave you time to reflect on the message you were sending. Many people today just sit down and write and send. This is a big mistake. People can interpret the message the wrong way. When you have a document or message, prepare it and save it. Let some time go by. Then go back and review what you have written—before you sent it. Check for understandability and look out for how the receiver might interpret what you are saying politically.

We suggested keeping project memos and notes brief. Another idea is to highlight your message in a bullet or outline form. Then write the text around the bullets or outline. Finally, keep in mind the type of writing you are doing. Is it marketing? Is it technical? Do not mix the approaches we discussed in an earlier chapter.

YOUR PRESENTATION SKILLS

Being on a project team means that you will have to make presentations. You will at least present material to the rest of the team and the project manager. Keep the message short and direct. People appreciate brevity. If you have to report on different issues during the project, establish a standard format and stick to it. People expect the same format so that they will concentrate on what you are saying as opposed to what the structure is.

If you are going to use graphs or tables, keep these to a minimum. You are trying to get a message across not trying to impress people with your ability to draw fancy charts and graphs. Spreadsheet and graphics software are getting easier to use and more powerful. It is tempting to make a presentation fancy. Then the presentation may cloud the meaning of the message. Using a few selected charts and graphs is a good idea. Many successful political campaigners have used charts and graphs effectively.

When you are going to make a presentation and have formulated a message, develop the minimum number of charts and graphs to make your point. Sit back and look at what else the graphs or charts tell. If they tell too much about some other topic, it may detract from your case. An old legal axiom is that one exhibit or chart in a case can serve as the point of focus and turning point in a case. You want the presentation to work in your favor, not against you.

KEY DECISIONS ON PROJECT MANAGEMENT

There are some critical decisions an organization must make regarding project management. The answers to the questions below determine how effectively the organization is using project management and what benefits it will derive from projects.

- Are projects to be managed separately in a stand-alone mode, or in a collaborative mode, as we have discussed, where resources are shared between projects?
- Will management actively direct multiple projects simultaneously as opposed to addressing individual projects sequentially?
- Will project management be a joint team activity in the organization, or be the activity of professional schedules and several project managers?
- Will information sharing among and within project teams be encouraged to address issues and shared project data (e.g., parts, vendors, etc.)?
- Will the organization actively embrace the gathering and dissemination of lessons learned from current and past projects?
- Will standardized high level templates be created and employed by all projects to ensure some level of compatibility?

SOME FINAL THOUGHTS

After reading this book, we hope you will be a better participant and manager of projects. In the 1990s, there are likely to be more projects. Projects give management accountability and tracability. They can watch what is going on more easily than if the traditional organization is performing the work. Projects and project management will become increasingly important. But they will tend to be smaller projects with focused goals and very specific, demanding schedules. Resources will be more limited. Thus, the need for effective project staff and managers will grow. Now it's your turn.

Index

The letters "ff" indicate the reader should also see the following pages.